The 90-Day

ISO

Manual
The Basics

James R. Stewart
Peter Mauch
Frank Straka

S^t_L

T5-AFS-386

TS
156
.S74X
1994
[V.1]

Copyright ©1994 by St. Lucie Press

All rights reserved. No part of this publication may be reproduced, stored in a retrieval system or transmitted in any form or by any means, electronic, mechanical, photocopying, recording or otherwise, without the prior written permission of the publisher.

ISBN 1-884015-11-5

Printed and bound in the U.S.A. Printed on acid-free paper.

All rights reserved. Authorization to photocopy items for internal or personal use, or the personal or internal use of specific clients, is granted by St. Lucie Press, provided that $.50 per page photocopied is paid directly to Copyright Clearance Center, 27 Congress Street, Salem, MA 01970 USA. The fee code for users of the Transactional Reporting Service is ISBN 1-884015-11-5 9/94/$100/$.50. The fee is subject to change without notice. For organizations that have been granted a photocopy license by the CCC, a separate system of payment has been arranged.

The copyright owner's consent does not extend to copying for general distribution, for promotion, for creating new works, or for resale. Specific permission must be obtained from St. Lucie Press for such copying.

Direct all inquiries to St. Lucie Press, Inc., 100 E. Linton Blvd., Suite 403B, Delray Beach, Florida 33483.

Phone: (407) 274-9906
Fax: (407) 274-9927

S^{t}L

Published by
St. Lucie Press
100 E. Linton Blvd., Suite 403B
Delray Beach, FL 33483

PREFACE

The subject of registering for ISO 9000 certification introduces many points of contention. Much misinformation currently exists in the industrial community. The motives of companies for obtaining registration, such as gaining market share or competing in Europe, are not the motives intended by the writers of the standard. Further, the registration process is complicated by firms more interested in obtaining clients than in ensuring the quality of U.S. products. Even at the certification level, political and organizational relationships have interfered with the recommended process of ISO 9000 implementation in business:

- Committing to operating a quality management system
- Installing the necessary components of such a system
- Considering the needs of the customer in the system
- Obtaining registration in accordance with the customer needs
- Continually improving the functioning system

This book is intended to help the quality professional install and operate a registered quality system, with appropriate assistance. It is also designed to aid the interested manager in understanding much of the important aspects of such a system.

At the time of this writing, the initial standard (ISO 9001 1987) is about to expire, and the revision is issued in draft (DIS-ISO 9001 1994). There may be substantial wording changes, but the authors do not feel that substantial philosophical or activity modifications will be made to the DIS copy. However, in keeping with the quality concept of being prepared for such contingencies, the implementation guide is presented in looseleaf format so that any such changes can be easily inserted.

Chapter 1 includes a discussion of the registration and certification process itself. The information was gathered from the two sources listed in the Endnotes, as well as from the authors' experience. The chapter illustrates the variety of politically and professionally oriented interactions that are often the result of ISO 9000 standards application, such as (1) the development of internal standards and company activities in order to comply with international standards, (2) the response of the various segments of the professional quality community to standards, (3) the activities of the federal regulatory bureaucracy in utilizing such standards, and (4) the international trade ramifications of the standard. Names and addresses of major organizations are also included. It is this chapter, if any, that may need to be updated with current information.

Chapter 2, Quality Concepts Contributing to the ISO Standard, is included because the authors consider themselves professionals involved in the continuous improvement of certifiable quality through the internal registration of ISO 9000. Compliance with this standard requires the adoption of a quality improvement philosophy by a company and its employees and the utilization of available quality tools. The registration process, by its nature, does not require any specific quality program. It does, however, require the existence of a program covering the twenty elements discussed in Chapter 2. The chapter provides a summary of the approaches used by major quality experts. It was developed from class notes and a number of books. At least one recent publication written by each of the recognized experts in the field is cited. It is the authors' intent that readers be able to identify appropriate approaches to their quality system problems within this book and locate any additional information necessary from the books listed in the Endnotes.

In Chapter 3, which is on applying the quality sciences, the approach taken is to look at the organization as a system and define how the ISO 9000 elements relate to such a system. It provides a connection between the abstract quality tools of the preceding chapter and the concrete registerable ISO 9000 system that a business desires. This chapter was drawn entirely from the consulting experiences of the authors. Many of the examples were developed from observation of successful and unsuccessful firms.

Chapter 4 covers total quality management (TQM). It is the belief of many experts, including the authors, that registration is only the beginning. As discussed in the chapter, Q94 is certainly a guide to a more complete TQM system and a goal for those who receive certification. Although there are a few organizations that will audit your system for compliance to Q94, it is not part of the registration scheme. It is expected that the Q94 will eventually become an effective measure of quality. Toward this end, recent work done by ANSI standards group Z-1 on expanding Q94 to reflect more complex TQM criteria has not been published. The authors provide the draft with comments about the post-ISO total quality system for those who desire a more in-depth understanding of quality.

The implementation guide, found in the looseleaf binder, consists of a series of implementation schemes that have been used by clients to meet the requirements of many registrars. Each scheme includes a purpose, some related background, an area of application, a description of a procedure, sample forms, and a description of how to complete such forms. All of the major topics are covered. Using these schemes, company quality personnel should be able to write a quality policy statement (for the quality improvement manual) and an appropriate procedure, design a form, and implement the scheme according to the company needs and culture.

It is schemes such as these that make the development of a 90-day ISO 9000 manual possible. It has been the experience of the authors that, utilizing these schemes and working closely with the employees of a company seeking ISO 9000, a registerable ISO system can be implemented within this time frame.

The ISO process has been considered too difficult by many companies. Plans costing hundreds of thousands of dollars and years of development are unnecessary, and may even be counterproductive, to meet the goal of registration. Unless an activity directly contributes to implementing or documenting one of the twenty elements, it is not effective in the ISO implementation strategy.

On the other hand, it is also the concern of the authors, as well as of many quality professionals, that inadequate compliance with the standard not become a quality disaster because a company provided only the minimal amount of reporting permitted by the ISO Audit System.

The implementation guide also includes a copy of the latest DIS draft of the ISO 9001 1994 Standard, in which the language has been clarified. Until its publication, the authors used a transition document (called the MBA ISO 9000 Standard) to explain what the standard really meant. This is no longer necessary. The document is included in the looseleaf section and the user can substitute the completed standard or future supplement if wording changes significantly.

The authors would like to hear of readers' experiences and suggestions about adding to the content of this book, particularly feedback about the use of a form or scheme mentioned in this book in a registration process and the response of the registrar. The authors will be glad to elaborate on implementation of the topics in the book.

James R. Stewart
Peter Mauch
Frank Straka

AUTHORS

Dr. James R. Stewart, PE, has over thirty years of experience as an engineering manager and educator and is an Assistant Professor at Northern Illinois University. He has developed Technology of Quality emphasis and currently teaches graduate classes in the Department of Technology on statistical process control, total quality management, design of experiments, and process analysis. He studied in England for his IQA certification (Ewbanks Preece) and is an advisor to a student chapter of the American Society for Quality Control.

Peter Mauch, CQA, CQE, is President of Mauch and Associates and has acted as an ISO 9000 consultant with over 35 firms. He teaches at several local community colleges. In addition, he is a member of the Technical Advisory Group 176 and has served as chairman of the Chicago section of the American Society for Quality Control.

Frank Straka, CQA, CQE, CRE, PE, has over 20 years of management experience in developing, implementing, and managing quality systems covering product development and production in service, manufacturing, information technology, software, electronics, and distribution industries. He is a member of ISO Technical Committee (TC) 176 (Quality) and TC 69 (Statistics) and IEC TC 56 (Dependability, previously called Reliability). He is also an executive board member of ANSI Z-1 committee for U.S. standards on quality, dependability, and statistics. He is currently an auditor for Underwriters Laboratories.

CONTENTS

CHAPTER 1

WHAT IS ISO 9000 AND WHY IS IT NEEDED?

It is apparent that many companies will either need to conform to the requirements of ISO 9000 or lose customers, even major blocks of business. It is useful to see how these standards came about because such information will be applicable as other such standards impact the American workplace.

DEVELOPMENT OF STANDARDS

How Was ISO 9000 Developed?

In 1959, the Department of Defense issued a number of product inspection standards. They included MIL-Q-105, defining acceptable quality limit (AQL) inspection sampling programs; MIL-Q-414, defining variables sampling; and MIL-Q-1235, defining continuous sampling plans. Standards such as MIL-Q-690 and MIL-Q-781 were developed for testing the failure rate and mean life reliability of products.

1

The time had come, however, for the suppliers of the U.S. military to adopt a quality systems approach. As a result, MIL-Q-9858 was written.

Subsequently, the North Atlantic Treaty Organization (NATO) adopted MIL-Q-9858 in 1969, after ten years of American revision and utilization. It was known as Allied Quality Assurance Publication 1 (AQAP-1). In the United Kingdom, the Department of Defence adopted the provisions of the standard as its Management Program Defence Standard DEF/STAN 05-8.

In 1979, the British Standards Institution adopted the military specification into the first commercial quality system standard, BS 5750. It was from this standard that the international committee chartered to write world standards developed the ISO 9000 series.

Since then, the British Department of Defence standard has been renumbered DEF/STAN 05-21, 22, 23, and 24 and revised to the ISO standard, and NATO is rewriting AQAP-1. Further, the U.S. Department of Defense is evaluating the use of the ISO 9000 series of standards.

It should be no surprise that the military continue to adapt and revise standards such as this one. However, it is the civilian aspect of this standard, combined with the increasing international market of U.S. industry, that is generating widespread interest. This standard, when adopted (with modifications) by the Automotive Industry Action Group (AIAG), will become the most significant supplier auditing system in the United States.

International Aspects of ISO 9000

International Organization for Standardization

The International Organization for Standardization (ISO) is a specialized international agency founded in 1946 to promote the development of international standards. Included in the scope of the agency is conformity assessment, including inspection, testing, laboratory accreditation, certification, quality system assessment, and other activities intended to assure product conformity. Currently, standardization efforts are directed by 91 standards bodies from as many countries, although not all are concerned with ISO 9000. The American National Standards Institute (ANSI) represents the United States. ISO is made up of 180 technical committees, each responsible for an area of specialization ranging from asbestos to zinc.

The object of ISO is to promote the development of standardization and related world activities in order to facilitate international exchange of goods and services and develop cooperation in intellectual, scientific, technological, and economic activities. The results of ISO technical work are published as international standards, and these are revised every five years.

The quality standards dealt with in this book comprise two series, ISO 9000 and ISO 10000. They are administered by one committee, Technical Committee 176.

Technical Committee 176

Technical Committee 176 (ISO/TC 176) was formed in 1979 to coordinate the increasing international activity in the area of quality management and quality assurance standards.

Subcommittee SC 1, Terminology, was established to agree on common definitions of quality terms. It developed ISO 8402: Quality–Vocabulary, published in 1986 (ANSI/ASQC A3-1987), which contained many of the terms and definitions from ISO 8402.

Also around this time, Subcommittee SC 2, Quality Systems, was established to develop quality systems standards, the end result of which was the ISO 9000 series (9000-1, 9001, 9002, 9003, 9004-1), which was published in 1987 and is currently under revision. The goal of Vision 2000, published by an ad hoc committee of TC 176, is to have a single total quality management standard by the year 2000.

Current standards, in draft or issued as additional parts, or sections, for ISO 9000 and ISO 9004 are as follows:

9000-2   Generic Guidelines for the Application of 9001, 9002, and 9003

9000-3   Guidelines for Software Development, Supply, and Maintenance (1991)

9000-4   Application for Dependability Management

9004-2	Guidelines for Services (1991)
9004-3	Guidelines for Processed Materials
9004-4	Guidelines for Quality Improvement
9004-5	Guide for Quality Assurance for Project Management
9004-6	Guidelines for Quality Plans
9004-7	Guidelines for Configuration Management

The third subcommittee, SC 3, Quality Technologies, has been charged with developing the standards for controlling the quality systems. In these standards are the rules for the auditing process. Standards issued or in draft from this committee include:

10011-1	Auditing
10011-2	Qualification Criteria for Auditors
10011-3	Managing Audit Programs
10012-1	Management of Measuring Equipment
10012-2	Measuring Assurance
10013	Guidelines for Developing Quality Manuals
10014	Guide to the Economic Effects of Quality
10015	Continuing Education and Training Guidelines

International standards and drafts can be purchased from:

The American National Standards Institute
11 West 42nd Street, 13th Floor
New York, NY 10036
Phone: (212) 642-4900

The ISO Forum

The ISO Forum was established by the ISO to disseminate information to users about ISO 9000 issues. Subscriptions are available in the United States through the American Society for Quality Control (ASQC). In addition to distributing the newsletter, the Forum has held symposia in a number of countries.

ISO Adoption

In addition to the adoption of the ISO 9000 series by the European Community (EC), over fifty countries have adopted and six are in the process of adopting the standard. As each nation adopts the standard, it assigns its own number and organizes the materials. The international number, U.S. number, and the originating country number are shown in the following table:

International	U.S.	Origin user
ISO 9000	ANSI/ASQC Q90	Canada (Z299)
ISO 9001	ANSI/ASQC Q91	U.K. (BS 5750 version of USA MIL-Q-9858)
ISO 9002	ANSI/ASQC Q92	U.K. (BS 5750 version of USA MIL-Q-9858)
ISO 9003	ANSI/ASQC Q93	U.K. (BS 5750 version of USA MIL-Q-9858)
ISO 9004	ANSI/ASQC Q94	U.S. (Z1)

Technical Advisory Group 176

The United States participates in the development process through membership in ISO via ANSI. This input is channeled through a Technical Advisory Group (TAG). ASQC administers, on behalf of ANSI, the U.S. TAG to ISO/TC 176, ISO 69, and IEC 56. Qualified U.S. experts participate in the meetings and work groups where documents are drafted. For information on TAG 176, contact ASQC at:

Standards Administrator
ASQC
611 E. Wisconsin Ave.
Milwaukee, WI 53202
Phone: (800) 248-1946

The U.S. Version of ISO 9000 Standards: Q90–94

In the United States, the ISO 9000 series is known as the ANSI/ ASQC Q90 series. These standards are technically equivalent to the ISO 9000 series and translate British usage to American usage. Future adoptions of ISO standards in the United States will be known as the Q9000 series. The ISO number will remain on the standard to reduce confusion.

The first standard in the series, Q90, is a road map to assist in the selection of three quality system models: Q91, 92, and 93. These models are actually successive subsets of each other. Q91 is the most comprehensive, covering design, manufacturing, installation, and servicing systems, whereas Q93 covers only final product inspection and test. The most common model, Q92, involves manufacturing for customer specification. All of the models were developed for use in contractual situations, such as those between customer and supplier. In 1987, the United States adopted the ISO 9000 series as the ANSI/ASQC Q90 series.

International	U.S.	Description	Use
ISO 9000	ANSI/ASQC Q90	Selection and use	Guide
ISO 9001	ANSI/ASQC Q91	Design/development, production, installation, and servicing	Contractual
ISO 9002	ANSI/ASQC Q92	Production and installation	Contractual
ISO 9003	ANSI/ASQC Q93	Final inspection and test	Contractual
ISO 9004	ANSI/ASQC Q94	Quality management and quality system	Guide

Elements covered in each of the three process standards are as follows:

Contractual elements	9001	9002	9003
4.1 Management Responsibility	Yes	Yes	Yes
4.2 Quality System	Yes	Yes	Yes
4.3 Contract Review	Yes	Yes	No
4.4 Design Control	Yes	No	No
4.5 Document Control	Yes	Yes	Yes
4.6 Purchasing	Yes	Yes	No
4.7 Purchaser Supplied Product	Yes	Yes	No
4.8 Product Identification and Traceability	Yes	Yes	Yes
4.9 Process Control	Yes	Yes	No
4.10 Inspection and Testing	Yes	Yes	Yes
4.11 Inspection, Measuring, and Test Equipment	Yes	Yes	Yes
4.12 Inspection Test Status	Yes	Yes	Yes
4.13 Control of Non-Conforming Product	Yes	Yes	Yes
4.14 Corrective Action	Yes	Yes	No
4.15 Handling, Storage, Packaging, and Delivery	Yes	Yes	Yes
4.16 Quality Records	Yes	Yes	Yes
4.17 Internal Quality Audits	Yes	Yes	No
4.18 Training	Yes	Yes	Yes
4.19 Servicing	Yes	No	No
4.20 Statistical Techniques	Yes	Yes	Yes

INTERNATIONAL STANDARDS APPLICATION

Regulation of Products for Sale in the European Community

For exporters of certain goods or services, called regulated products (see following list), specific product certification requirements must be met in order to sell in the European market. The ISO 9000 standards are only a portion of such requirements. Each of the classes has its own directive and requirements, usually ISO 9000 in addition to a product testing plan. Although the ISO 9000 standards are *not* product standards, most products require quality system approval as a part of product certification. The certification requirements, like many trade agreements, are quite complex, and alternate methods of compliance may be available that do not require quality system approval.

Manufacturers of regulated products must be certain to comply with the most stringent of all relevant requirements. Each directive may require different methods of proving conformity. It is entirely possible that a company may meet the requirements for ISO 9000 and also have product that meets one of the conformity options, but find it may not be exportable to the EC.

Consideration of options by a company planning to export into the EC requires an additional decision. In certain cases, ISO 9000 may be one of three options. The compliance directive may specify that the quality system requirement may be completed without any ISO 9000 consideration, by self (first person) audit, by a supplier (second party) audit, or by a third-party audit conducted by a specific agency only. Some examples of regulated products are as follows:

Elevators

Cableways equipment

Construction products

Medical devices

Implantable medical devices

Personal protective equipment

Telecommunications equipment

Gas appliances

Nonautomatic weighing instruments

Pressure equipment

Measuring and testing instruments

Equipment in explosive atmospheres

Furniture (flammability)

ISO/IEC Guide 25

The International Electrotechnical Commission (IEC), in conjunction with ISO, has adopted Guide 25, General Requirements for the Competence of Calibration and Testing Laboratories. The guide applies to labs acting as suppliers that produce calibration and test results, other pertinent clauses, and certain technical requirements such as technical competence of laboratory personnel, adherence to specific test methodologies, and participation in proficiency testing. There is considerable similarity, or perhaps overlap, in the scope of the two standards. Organizations may be accredited to this guide in lieu of ISO 9000.

European Community
Product Safety and Liability Directives

In 1985, the EC Council issued Directive 85/374, known as the Product Liability Directive. It defines product liability and introduces a uniform concept of such liability throughout the EC. Products are considered to be defective when they do not provide the amount of safety the public has the right to expect. An injured customer must show damage experienced, the product defect responsible for the problem, and the relationship between them. It is not necessary to prove that the product is unreasonably defective, and if the manufacturer could have reasonably foreseen the problem, negligence does not have to be proven. Other directives, such as Medical Device and Machine Safety, may define a more precise level of safety.

Directive 92/59, the Product Safety Directive, covered all products in the EC market when it became fully implemented in June 1994. Products intended for customers must not present any unacceptable risks, and potential users of such products have to be adequately warned of any remaining risks. It is intended to impose a general criterion on producers, requiring them to introduce only safe products. The safe product directive requires monitoring of direct or indirect safety or health risks of the product, alone or in combination with other products, by each EC country over the life of the products, including design, composition, execution, functioning, wrapping, conditions of assembly, maintenance or disposal, instructions for handling and use, and any other of the properties of the product.

In the event of product safety liability claims, registration to ISO 9000 alone does not provide a defense. However, some legal opinion has suggested that such registration, when combined with both product safety technical documentation and adequate labeling and user instructions, could prove useful.

Quality System Registration

Even if a company does not manufacture a product listed in the regulated lists, marketplace demands rather than regulatory authority may require ISO 9000 registration. Procurement authorities or buyers may simply require that companies be audited and registered as in compliance with an ISO 9000 standard. Such purchasing requirements are especially likely to apply in industries such as automobiles, electronics, aircraft, measuring and testing equipment, machine tools, health care products, and other industries where there are safety and liability concerns. Vendors in such industries desiring to sell to companies imposing ISO 9000 registration requirements will need to be audited and registered as being compliant under terms acceptable to the purchasers.

European Accreditation of Certification (EAC)

In 1991, a memorandum of understanding (MOU) was signed by 13 nations: Belgium, Denmark, Ireland, the Netherlands, United Kingdom, Germany, Greece, Italy, Portugal, Iceland, Norway, Sweden, and Switzerland. The purpose of the organization was to create a single European system for recognizing certification and registration bodies that would provide adequate assurance that the process would be equivalent in all European countries.

The five goals of the EAC include:

• Maintain and strengthen market confidence in certificates issued by accredited bodies

• Establish mutual confidence among participating bodies and promote collaboration and agreements as a means toward a European system of assessment and accreditation

- Provide the means for a continuous flow of knowledge relevant to assessment and accreditation

- Work towards a multilateral agreement on the equivalence of the operation of the participating bodies and a declaration of their commitment to foster general acceptance of the equivalence of certificates

- Promote the harmonization of the operations of participating bodies

European Organization for Testing and Certification (EOTC)

In April 1990, the EC entered into a memorandum with major European Free Trade Association (EFTA) partners: Austria, Finland, Iceland, Liechtenstein, Norway, Sweden, and Switzerland. Included in the agreement were the two European standards development associations: the European Committee for Standardization (CEN) and the European Committee for Electrotechnical Standardization (CENLEC).

The organization thus formed promoted mutual recognition of test results, certification procedures, and quality system assessments and registrations in nonregulated product groups throughout the EC and EFTA countries. The EOTC also became responsible for providing technical assistance to the EC in preparation of mutual recognition agreements and other documents. Recognizing the need for expert advice on quality assurance matters, the EOTC has invited two organizations to its meetings, the European Organization for Quality (EOQ) and the European Committee for Testing and Certification (EQS).

The EOTC can be contacted at the following address:

European Organization for Testing and Certification
Rue Stassart 33
2nd floor, B-1050
Brussels, Belgium
Phone: 32 2 519 6969

The European Organization for Quality (EOQ)

The EOQ is a European organization whose mission is to improve the quality and reliability of goods and services, principally through publications and training. The ASQC is an affiliate member. The EOQ acts as an observer to the EOTC.

The European Committee for
Quality Systems and Certification (EQS)

With the goal of avoiding multiple quality system assessments and registrations of companies, this committee is responsible for collating the rules pertaining to the two areas. It utilizes the EC standard EN 45012, General Criteria for Certification Bodies Operating Quality System Certification, and promotes the blending of rules and procedures used for quality system assessment and registration among member nations.

The committee is an observer to the EOTC and is a candidate to become a functional committee of the EC.

European Network for
Quality System Assessment and Certification (EQNET)

Eight European quality system registration bodies (one not-for-profit organization from each country) have entered into a business agreement to establish close cooperation leading to mutual recognition of registration certificates. The members are AFQ (France), AIB-Vincotte (Belgium), BSI Quality Assurance (United Kingdom), DQS (Germany), DS (Denmark), N.V. KEMA (the Netherlands), SIS (Sweden), and SQS (Switzerland).

Each of the member companies agrees to:

- Promote the recognition of each other's quality system registration certificates
- Coordinate the work to be performed for quality systems registration of an organization having subsidiaries in several EC/EFTA countries, in order to help such an organization to obtain appropriate quality system (QS) certificates

- Issue several QS certificates simultaneously after performance of a joint audit
- Promote bilateral agreements between EQNET members
- Contribute to the development of operating procedures and promotional materials
- Present information on EQNET

Agreements between the EC and Other European Countries

The twelve EC and seven EFTA countries have entered into bilateral association agreements with the rest of the European nations. Poland, Hungary, Albania, and the Baltic States have had agreements for several years. Others, such as Bulgaria, Romania, and the former Soviet states have recent agreements or are in negotiation. These agreements will gradually establish bilateral free trade, with the ultimate goal of EC membership.

The ISO 9000 Standard and Non-European Countries

Just as manufacturers in the United States are adopting the ISO 9000 standard in order to compete in the European community, other nations are also actively involved (Peach, 1992). Future trade agreements in Southeast Asia may include ISO 9000 simply because there are national standards in Australia, China, India, Japan, Malaysia, New Zealand, Pakistan, Philippines, Singapore, South Korea, and Thailand.

All members of NAFTA have adopted the standard, as well as Argentina, Brazil, Chile, Colombia, Venezuela, Cuba, and Jamaica.

ISO 9000 Standards for the United States

There are three areas of growth for the ISO 9000 standards: in regulated products for sale in the EC, in government procurement, and in nonregulated products. The certifying agencies involved, registration procedure, and requirements may be quite different depending on the company's marketing plan.

Regulated Products

All international trade agreements, including quality system agreements, are made through the U.S. Department of Commerce. Within the Department of Commerce, one program covers the EC single market. Information can be obtained about EC regulations, background of the EC, or assistance with specific trade opportunities or problems from:

The Office of European Community Affairs
International Trade Administration, Room 3036
14th and Constitution Ave., NW
Washington, DC 20230
Phone: (202) 482-5276

In an agreement (Breitenberg, 1993) between the EC and the U.S. Department of Commerce, the National Institute of Standards and Technology (NIST) was proposed as the agency to provide the EC with necessary assurances about the competence of U.S.-based testing, certification, and quality system registration bodies to conduct conformity assessment activities under U.S.–EC MOUs. The program, developed by NIST, is called NVCASE and includes the following characteristics:

- Recognizes qualified conformity assessment to gain greater acceptance of U.S. products
- Is voluntary (no organization is required to apply)
- Is a self-supporting and fee-for-service program
- Is limited to areas of industry-required conformity assessment
- Operates in product areas regulated in foreign countries
- Operates at the level of accreditation for certification, laboratory accreditation, and quality system registration
- Does not act as a certifier, testing laboratory, or registrar
- Does not operate in product areas covered by other federal agencies unless requested by that agency

Information on U.S., foreign, and international voluntary standards can be obtained from:

National Center for Standards and Certification Information
National Institute of Standards and Technology
TRF Bldg. A163
Gaithersburg, MD 20899
Phone: (301) 975-4040

Government Procurement

The Department of Defense and the military services (Breitenberg, 1993) will probably adopt ISO 9000 as the standard for quality management. As a result, MIL-Q-9858A and MIL-I-45208 will be gradually phased out. Registration requirements will be replaced by second-party inspection or specific quality system requirements. When a product certification is required, a new MIL-Q-9858B will be implemented.

The Food and Drug Administration is in the process of adopting ISO 9000 under a four-nation agreement to replace the current Good Manufacturing Practices (GMP) standards for medical devices. This is a complex process, but the United States will ultimately be compatible with the EC, Canada, and Japan.

The Nuclear Regulatory Commission is prepared to apply ISO 9000 as an option for its regulatory reviews. Such audits are still conducted according to 10 CFR 50 standards.

Non-Regulated Products

The accreditation of registrars in the United States is quite complex. Further, the selection of a registrar may have far-reaching ramifications in the future. A company may very well go through an expensive registration process only to find that the qualification for a registrar does not meet the requirements of the company's customer.

In the United States, the Registrar Accreditation Board (RAB), a joint venture of ASQC and ANSI, is the agency for evaluating the quality of services offered by registrars. A process that meets the three parts of ISO 10011 and follows the ISO Guide 48 has been established for certifying registrars and their auditors and for recognizing auditor training courses. Certification of registrars started off slowly. For the first several years, only five companies were certified as registrars. Further, there have been questions about the transferability of the registrar certificate. As RAB increases its influence on the registrar certification process and negotiates an appropriate number of MOUs with other certifying agencies, the system may become more functional. RAB can be contacted at:

Registrar Accreditation Board
611 East Wisconsin Ave.
PO Box 3005
Milwaukee, WI 53202
Phone: (414) 272-8575

Registrars in the United States may alternatively certify with the Dutch Council for Accreditation (RvC). It has accredited registrars in several non-EC countries as well as in the United States. RAB has completed a MOU with the RvC. The Standards Council of Canada (SCC) also has a registration open to U.S. registrars.

In addition, some U.S. registrars have parent companies in European countries certified through their home countries.

Benefits vs. Cost of ISO 9000

There are several benefits to implementing ISO 9000 quality system assessment standards in a company. For example, it will provide a guide for building quality into the product or service. In addition, the number of audits customers perform on the operations may be reduced. Increasingly, customers are accepting supplier certification from an accredited third-party assessment company based on these standards.

Corporations around the world have been building, and continue to build, their quality systems based on the ISO 9000 standards. Both large and small companies with international business perceive the ISO 9000 series as a route to open markets and improved competitiveness.

Of all the questions about the ISO 9000 series, those concerning registration cause the greatest concern. It is important to be aware that the ISO standards may be used as an internal company goal, without having to take the further step of obtaining registration. In addition, they may be used as a procedure for company personnel required to audit suppliers. However, without the benefit of a quality professional experienced in manufacturing a company's product with an unbiased view of the business, the company will not get the full benefit of the standard. Very few companies would conduct their financial affairs based solely on internal audits.

Increasingly, European customers expect U.S. companies to have their quality systems registered to ISO 9001, 9002, or 9003. This generally involves having an accredited third party conduct an on-site audit of the company operations against the requirements of the appropriate standard. Once a company has been registered, it will receive a registration certificate to one of the ISO 9000 standards. It will also be listed in a register maintained by the accredited third-party registration organization. During the registration period, the registrar may conduct periodic surveyance audits. The company is expected to operate according to the quality plan that has been installed. For maximum impact on their customer base, many companies publicize their registration with the appropriate ISO standard.

Some of the benefits of registration are as follows:

- Verifies a commitment to quality
- States that a level of performance has successfully been established
- Demonstrates a customer focus
- Renders quality audits unnecessary
- Improved product/service
- Increased positive image of company
- Enhanced competitive position
- Reduced costs

Preparing for an ISO 9000 quality system audit will mean that a careful analysis of all sources of nonconformities must be conducted and methods for reducing any nonconformities be found. It may be that a company has none of these improvement opportunities, in which case ISO 9000 certification should be quite easy. Typically, it has been found that when the costs of the registration process and the benefits of the improved operation are compared, approximately a 5-to-1 return on investment is realized. The following are examples of activities where savings were found after ISO 9000 registration:

Activity	Description of saving
Sales order review	Reduction of errors in sales order taking
Business planning	Insurance that all members of the organization understand what they are expected to do
Production control	Reduction in customer shipping errors
Inspection and testing	Reduction in inspection and test errors
Purchasing	Reduction in supplier defects and late deliveries
Packaging and storage	Reduction of errors made in product packaging and material storage

Selecting the Registrar

As previously noted, selection of an inappropriate registrar can cause considerable problems in meeting the company's goals for acquiring ISO 9000 (Breitenberg, 1993). There are, however, a number of questions that should be asked before deciding on a registrar. It is important to the registrar that the company be happy with the service, and it is certainly important to the company that it receive the appropriate service. Questions to ask include:

- Are there regulatory requirements or marketplace requirements for the company product?
- Does the registrar meet these requirements?
- Which registrars do the customers prefer and accept?
- What is the scope of the registrar's accreditation?
- What MOUs does the registrar have?
- Do the MOUs match the company marketing plan?
- What references does the registrar supply?
- How does the registrar protect confidential information?

- What is the appeals process for complaints?
- What are the policies in the event of suspension?
- Are registrar responses timely and appropriate?
- Are system deficiencies adequately documented?
- How soon can the registration be completed?
- Does the registrar regularly update a list of registered companies?
- What are the skills, training, and experience of the auditors?
- Does the company have the right to review the auditor profile for acceptance?
- What is the accreditation of the auditors?
- Are the auditors quality professionals?
- In general, does the registrar comply with ISO Guide 48?
- What are the services (process) in certification?
- What are the costs of certification?

Costs of Registration

The costs of the audit and registration should not be a major factor for most companies. The internal costs of preparing for the audit will be much higher. Typically, a company will pay approximately $1300 per person per day for an audit. Total registrar costs approximated for a three-year period are as follows:

ISO 9001	$12,000 to 15,000
ISO 9002	$8,000 to 12,000
ISO 9003	$2,000 to 3,000

These costs will vary with the complexity of the system, number of people, and scope of the registration.

Services provided by the registrar will vary slightly, but generally include documentation review, audit, and annual follow-up audits during a three-year period. A preliminary audit may be added. The costs of any follow-up due to nonconformities are at the request of the client. One schedule of the steps for a typical registration is as follows:

Activity	Description
1.0 Company	Requests details of registration
2.0 Registrar	Customer relation discusses all aspects of the registration program
3.0 Registrar	Prepares proposal and mails it to company
4.0 Company	Reviews proposal, completes and signs application form, and returns to registrar
5.0 Registrar	File number given to company; assigns auditor and informs company
6.0 Auditor	Explains how registration works and provides process information
7.0 Company	Submits company Quality Policy Manual
8.0 Auditor	Reviews manual
9.0 Company	Revises manual as necessary
10.0 Auditor	Reviews revisions
11.0 Auditor and Company	Establish date for audit
12.0 Auditor and Company	Audit of company facility
13.0 Auditor	Preparation of registration documentation
14.0 Review Board	Reviews registration documentation, approves registration of company
15.0 Registrar	Congratulatory letter, registration certificate
16.0 Registrar	Compliance audit first year
17.0 Registrar	Compliance audit (2) second year
18.0 Registrar	Compliance audit (2) third year
19.0 Registrar	Re-registration (see step 3.0)

The Role of Consultants in ISO 9000 Registration

The document by which all EC and other nations intending to use the standard have agreed upon, ISO/IEC Guide 48, states in part:

> An organization that, directly or through the agency of sub-contractors, advises a company how to set up its quality system, or writes its documentation **should not provide assessment services** to that company, unless strict separation is achieved to ensure that there is no conflict of interest.

The rule against assessment (consulting) by registrars is necessary because the audit is viewed as an impartial assessment of the company's system against the standard. The auditor is not permitted to act as a consultant (i.e., explain how to design the manufacturing system to meet the requirements), but can only assess whether or not the requirements have been met. During an audit, it is vital that someone be present who understands the significance of the comments and nonconformities the auditor may note.

While any company with a comprehensive quality program can document the process sufficiently to pass ISO 9000 requirements, the direction and spirit may need to be developed by quality specialists. For obtaining registration, the tasks are quite different from the registrar's function. If the consulting is done internally, the person must be able and willing to:

- Look at all areas of the company and ascertain nonconformities
- Suggest and implement procedures to eliminate such nonconformities
- Evaluate audit reports from the registrar
- Interface with the organization pursuant to modifying the process in accordance with registrar findings
- Respond appropriately to arbitrary audit findings

If a consultant is deemed appropriate for the company, two things should be ascertained:

- The consultant is conversant with ISO 9000 applications to the particular industry. The consultants should be ASQC certified as a Certified Quality Auditor (CQE) and Certified Quality Engineer (CQA), have experience in quality management, and be knowledgeable about ISO.

- The consultant is providing those services necessary for ISO certification. Some consultants would like a company to be an excellent example of some TQM program. This is neither necessary nor desirable. Experience with implementing quality systems encompassing each of the twenty elements is more useful. Other consultants have a hidden agenda. Blueprint reading, literacy, or statistical process control (SPC) are *not* required for registration. Other consultants insist that inspection and sampling are the only acceptable statistical techniques. All of these approaches will cost time and money without moving the company toward certification. However, if an organization is serious about quality, it will need to meet the intent of ISO, and not just the wording.

ENDNOTES

Breitenberg, M. (1993). *Questions and Answers on Quality, The ISO Standards Series, Quality Systems Registration, and Related Issues.* U.S. Department of Commerce: Gaithersberg, Md., pp. 10, 13, 14.

Peach, R.W. (1992). *ISO 9000 Handbook.* CEEM: Arlington, Va., p. 4.

CHAPTER 2

QUALITY CONCEPTS CONTRIBUTING TO THE ISO STANDARD

HISTORY OF QUALITY CONTROL SYSTEMS

The ISO series of standards provides the framework necessary for the knowledgeable outsider—whether from a different department, customer company, or third-party assessment group—to use as a means of evaluating the quality efforts of an organization. The development of these standards is the result of an evolution in management attitudes and practices toward defining and achieving customer satisfaction. The evolution of quality control is explored in this chapter. Then some of the concepts used by major total quality management implementers are described and related to ISO 9000.

Quality in a Craft Society

Before the Industrial Revolution, craftsmen almost exclusively produced items that were provided directly to consumers. Handcrafted items for the common man were required to wear well or the maker would be shunned. Those who dared to sell their goods to the nobility had to meet the highest standards, for the consequences of inattention to detail were more severe. Many abbeys and castles had dungeons to serve that very purpose.

As the replacement of manual labor with mechanical devices became widespread, the nature of labor changed. In the past, a person did a complete task under contract. The mill employee used the machine to complete only that portion of the task for which the particular machine could be harnessed. The foundry, turning, boring, and assembly operations were each done in their own environment in the mill. The organization that provided the orders, supplied materials, and removed completed units became structured around these mill locations. Quickly (probably as soon as a foreman received unusable goods from another foreman), inspection became one of the activities in the plant. Just as the need for customs inspectors arose at national borders as the great nations arose, the need for product inspectors grew as the great mills were built.

Quality in the Mill

As production systems became more organized and the mill became the assembly factory, the job of quality became recognized, and persons were assigned to look for the defects in production. To fulfill his responsibility toward quality, the early foreman simply assigned those persons who could not complete a "fair day's work at the line" (older or weaker persons or the temporarily infirm) the task of inspection. Training was not required because such persons worked in the shop and knew what was right and what needed rework. In spite of the available expertise, the customer frequently received products that should have been reworked.

The change in the level of sophistication of industrial organization that occurred as a result of preparation for World War I and, in particular, in the pioneering industrial engineering work of Gantt and the Gilbreths led to the development of process sheets, routing documents, and schedules. It formalized the flow of production and provided a means for increasing the output and reducing the production of defects. A place for the inspector, the Quality Department, was located in each factory. To complete inspection of the product, the new managers turned to the task and again recruited those who knew it best—the misfits such as the old, weak, and infirm—who were happily transferred by their former foremen. Even with an emphasis from higher up in the company, the attitude that prevailed was that the quality department was to prevent defects from being shipped while supporting the production operation in reaching its output goals. Co-operation was clearly expected from the foremen, but it was still not uncommon for the customer to receive unusable materials. The separate quality department, although it did not solve the fundamental goal of eliminating customer dissatisfaction, did result in placing the spotlight on waste and waste costs.

Beginnings of Quality Control

An obscure group at Bell Telephone Laboratories began operation in 1921. It was headed by Walter Shewhart, who in 1932 published *Economic Control of Quality of Manufactured Products* (the first book on quality), and included men such as Harold Dodge and Harry Romig, who jointly published the first sampling tables under the descriptive title, "Sampling Inspection Tables." They developed the techniques of sampling and control that are collectively labeled Statistical Quality Control. When the *Bell System Technical Journal* appeared in 1922, it contained the first papers on inspection and quality control. Throughout the 1930s, the small group at the Bell System attempted to improve the reliability of mechanical switching circuits. For years, however, the only response to quality was rejection of papers and presentations to sparse audiences. Quality topics were considered to be too cumbersome and complicated, too costly, and too academic.

As the nation grew in its role as the "arsenal of democracy" and had to provide a manufacturing base necessary to sustain the Allies in World War II, the need for eliminating strategic materials waste and creating consumer-acceptable devices became apparent. There were planes that couldn't fly, tank engines that wouldn't run, trucks that wouldn't back up, and bombs that wouldn't explode. Even worse were those products (such as torpedoes) that would return to the launcher and work all too efficiently.

In response to the need for quality, the first standard, MIL Std 105, was developed. In 1942, training courses were organized by Eugene

Grant, Holbrook Working, and W. Edwards Deming at Stanford University. The few Shewhart disciples then taught 31,000 persons the new statistical quality control in ten-day courses. These courses, sponsored by the U.S. Office of Education, trained government employees, managers, and many others. They were, however, directed toward supervisors—thousands of them—enthusiastic, young, and eager organizers of the war effort. The American Society for Quality Control was founded in 1946 as a result of this effort.

Although the results of the training continued into peacetime, its momentum diminished as management emphasis petered out. Quantity—getting the numbers out—replaced any focus on quality. American managers of the 1950s, for the most part, consisted of those who considered the Shewhart methods time-consuming and unnecessary. Although by 1949 Dr. Deming felt that "there was nothing—not even smoke," the outlook was not that bleak. Experts such as Joe Juran, Leonard Seder, Dorian Shainin, Warren Jones, and even Deming developed new techniques. Even industry half-heartedly accepted the need for quality.

The Quality Improvement Era

Although there was little interest from U.S. management, the rebuilders of the new industrial base listened intently. Many experts, including Dr. Deming and Dr. Juran, went to Japan to assist the reconstruction.

These expatriate experts taught that quality improvement was a systematic process. Although the various approaches were different, the end result—improved quality and profitability—was the same. Japanese statisticians eventually would redefine the manufacturing process, and Japan's war-torn industries were rebuilt with the new guidelines. Total Quality Management (TQM), a term coined in the United States by Armand Feigenbaum while he was a doctoral student at the Massachusetts Institute of Technology, defined the new quality system and became, along with the company song, the daily anthem of the Japanese corporation. Quality improvement in the world-class corporation has become the driving force for changes in market research toward the servicing and disposal of products.

Implementing TQM spread quality craftsmanship throughout the organization, thereby creating a need for some method for evaluating the effects of individual effort. Thus, auditing procedures were developed in the TQM system, as described by both Dr. Feigenbaum and Dr. Ishikawa. In addition, the ISO 9000 family of standards, along with the Japanese Deming Prize, the Baldrige Medal, the European Medal, and the prizes given by various American states, all reflect auditing application.

While prizes and medals attempt to measure the companies providing the highest level of quality through customer service, ISO 9000 standards provide a way of auditing existing quality practices in a company, assuring customers that a comprehensive quality system exists.

ASPECTS OF ISO 9000 IMPLEMENTATION

Quality Cost

Philip Crosby, another quality pioneer, relies on organizational and management processes to change corporate attitudes. Three key elements in his philosophy are determination, education, and implementation. Determination to have a successful quality program is required of top management, education in the fundamentals is required of everyone in the company, and implementation must be understood by the management team.

Crosby's standard for the organization is *zero defects* or, in other words, error-free work. He believes that people are conditioned to accept as inevitable flawed manufacturing, although in most areas of private life such error is not accepted. By accepting that error is inevitable, people allow a lack of attention. It is this lack of attention that causes errors. In his opinion, therefore, only error-free work is acceptable as a standard.

Quality is defined as conformance to requirements. Set by management, requirements—clearly stated so there can be no misunderstanding—are ironclad devices for communicating need. If these requirements are conformed to, there is no quality nonconformity.

Problems are never merely quality problems. Every nonconformity should be classified, reported, and directed to the individual or department that caused it. Appropriate actions should be taken by these departments.

Cost of Quality

Crosby (1979) was one of the first persons to measure performance of a quality system by segregating quality costs. Costs, under a Crosby system, are divided into the appraisal, prevention, and failure categories. In his system, doing it right yields the lowest cost. All other costs develop from correcting actions not done properly. His slogan, perhaps his epitaph—"quality is free"—is especially insightful when one recognizes how much costs are associated with finding or preventing nonquality.

Management, observes Crosby, reviews the costs of inventory, employee compensation, and sales projections. Why not review the costs of quality? Historically, quality has suffered from the lack of an obvious method of measurement.

The cost of quality analysis is determined by developing the fully loaded costs of the easy-to-gather information. This includes only one third of the total costs, but such rough estimates can show that cost of quality actually provides an improvement opportunity. The information to be gathered includes seven items:

1. Costs of doing things over, including clerical tasks
2. Costs of scrap
3. Costs of warranty and returns
4. Costs of after-service warranties
5. Costs of complaints
6. Costs ofinspection and tests
7. Costs of miscellaneous documents, such as change orders

Once an approximation of the total cost is known, plans for reduction can be made. Then, the costs can be sorted into three categories:

prevention, appraisal, and failure. A company should not wait for a full, formal system to be checked out before beginning a program of cost reduction.

Implementing the System

As the system is put into service, the cost of prevention is expected to grow into its major cost component. Crosby contends that this cost mitigates much higher failure costs.

Improvements to the quality system are achieved by implementing four programs. In the first, the maturity grid, the present system is evaluated and areas for improvement are measured. A second program, the fourteen-step quality improvement, implements quality practices for manufacturing. The third, the make-certain program, pertains to defect prevention for white-collar employees. Finally, modifications to the personal management style of decision-makers provide the fourth component of a Crosby system.

The Management Maturity Grid

As has been stated, one of the four programs is the process that Crosby (1979) provides to analyze an organization. The maturity grid (see Figure 2.1) can be utilized by any outside observer. Six measurement categories are ranked into five progressive stages: uncertainty, awakening, enlightenment, wisdom, and certainty. As the growth occurs, management becomes more capable of recognizing and solving quality problems. The levels between certainty and uncertainty are defined according to the organizational understanding and correction of problems.

Uncertainty represents a lack of knowledge of how problems can be reduced. The management simplistically views quality as a tool to prevent the error-makers from ruining the company. The quality "policemen" are often placed low in the organizational hierarchy. Every problem, even recurring ones, is treated as unique. Identifying who caused the problem is the major action. Uncertainty-ridden companies replace solving problems with working hard.

QUALITY MANAGEMENT MATURITY GRID

Rater _____

Measurement categories	Stage I: Uncertainty	Stage II: Awakening
Management understanding and attitude	No comprehension of quality as a management tool. Tend to blame quality department for "quality problems."	Recognizing that quality management may be of value but not willing to provide money or time to make it all happen.
Quality organization status	Quality is hidden in manufacturing or engineering departments. Inspection probably not part of organization. Emphasis on appraisal and sorting.	A stronger quality leader is appointed but main emphasis is still on appraisal and moving the product. Still part of manufacturing or other.
Problem handling	Problems are fought as they occur; no resolution; inadequate definition; lots of yelling and accusations.	Teams are set up to attack major problems. Long-range solutions are not solicited.
Cost of quality as % of sales	Reported: unknown Actual: 20%	Reported: 3% Actual: 18%
Quality improvement actions	No organized activities. No understanding of such activities.	Trying obvious "motivational" short-range efforts.
Summation of company quality posture	"We don't know why we have problems with quality."	"Is it absolutely necessary to always have problems with quality?"

Figure 2.1 Maturity Grid (*Source:* Crosby, P. (1979). *Quality Is Free.* Mentor: New York, pp. 32–33. Reproduced with permission.

Unit _____

Stage III: Enlightenment	Stage IV: Wisdom	Stage V: Certainty
While going through quality improvement program learn more about quality management; becoming supportive and helpful.	Participating. Understand absolutes of quality management. Recognize their personal role in continuing emphasis.	Consider quality management an essential part of company system.
Quality department reports to top management, all appraisal is incorporated and manager has role in management of company.	Quality manager is an officer of company; effective status reporting and preventive action. Involved with consumer affairs and special assignments.	Quality manager on board of directors. Prevention is main concern. Quality is a thought leader.
Corrective action communication established. Problems are faced openly and resolved in an orderly way.	Problems are identified early in their development. All functions are open to suggestion and improvement.	Except in the most unusual cases, problems are prevented.
Reported: 8% Actual: 12%	Reported: 6.5% Actual: 8%	Reported: 2.5% Actual: 2.5%
Implementation of the 14-step program with thorough under-standing and establishment of each step.	Continuing the 14-step program and starting Make Certain.	Quality improvement is a normal and continued activity.
"Through management commitment and quality improvement we are identifying and resolving our problems."	"Defect prevention is a routine part of our operation."	"We know why we do not have problems with quality."

Awakening recognizes that quality problems often recur and that some program may help eliminate such problems. However, in this case, resources are not available. Fads and piecemeal approaches allow a glimmer of what could be attained, but they often fail as soon as the novelty wears off. Costs are unavailable to justify a comprehensive program. Overall, the quality improvement function is disorganized, and results are transient.

Enlightenment occurs when a formal, company-wide improvement program is instituted for the implementation of a quality policy. The cost data that are available are accurate enough to make some estimates in establishing priorities and recognizing effective actions. Here, the quality department is developed into a well-balanced, functioning unit. The quality team approaches problems openly, searching for solutions rather than blame. However, until the next stage is reached, quality still emphasizes current, immediate problems rather than prevention of future problems.

Wisdom is the result of incorporating the quality programs into the organizational system, where they take hold and become a permanent part of company operations. Quality is given status. The quality manager may even be promoted to vice-president. Employees wonder how they used to handle all the problems. Attitude, systems, and enthusiasm are available. The quality system is real.

Certainty exists after the prevention system matures enough so that problems rarely occur and those that do are rapidly resolved. With the emphasis on quality prevention, the organization knows "why it does not have problems with quality."

Crosby's Quality Improvement Program

The fourteen-step program to measure and improve quality can be summarized as follows:

1. Begin with management commitment. The management position must be expressed clearly, both in activity and in communication. Crosby suggests a statement such as:
   Each individual is expected to perform exactly like the re-

quirement or cause the requirement to be officially changed to what we and the customer really need.

Management must recognize that a personal commitment is essential to ensure active cooperation.

2. Bring together people who can commit to action, preferably department managers. Orient employees and provide training.

All the resources necessary to do the job should be grouped in one team.

3. Develop measures for manufacturing and nonmanufacturing activities. Record status and corrective action. Place the results in a highly visible format, such as posted charts. Teach employees how to understand the charts.

A company-wide measurement system establishes the foundation of the quality improvement system.

4. Estimate the cost of quality. Establish the definitions for the elements in the cost of quality. Indicate corrective actions. Utilize the comptroller's office.

The company's system will contain a quality performance measure.

5. Share the costs of nonquality with the employees. Train supervisors to train the employees. Provide visual materials. It is a sharing rather than a motivating process.

Supervisors and employees will talk positively about quality.

6. Corrective action within the department is begun. Building on the habit of identifying and correcting problems, solutions are found or the problem is sent to, and resolved at, the next higher level.

The habit of identifying and resolving problems is begun.

7. The concept of zero defects (do things right the first time) is introduced to a management team. After adapting the concept to the company culture, the team should be prepared to implement the program.

As the general commitment runs out, a new commitment to a specific goal is desirable to maintain group cohesiveness.

8. Formal training is prescribed for all supervisors so they can explain each step to their people. Supervisors must know the fundamentals of zero defects and the details of the error-cause-removal process that accompanies a zero defects program.

9. A Zero Defects Day should be held to establish zero defects as the standard of the company. At the end of the day, employees understand their responsibilities in the same way and the day-long program emphasizes the subject and provides a long-lasting memory.

10. Specific, measurable goals are established. Usually, they are to be completed in less than 90 days.
    Meeting goals through teamwork is begun in the company.

11. Error causes are actively sought out and eliminated. This is not a suggestion system. Individuals describe problems in their workplace and appropriate professionals develop answers in a reasonable time.

12. Employees and groups that meet their goals or perform outstanding acts must be given recognition. Rewards should not be financial. It is wise not to attach relative values to the identification of problems, but genuine recognition is something all appreciate.

13. Quality councils are established to communicate and determine actions to improve the program in the future. They will be the best source of information on program status.

14. The last of the fourteen points is to do it over. Repetition makes the program perpetual. Crosby suggests that the improvement team change and that areas of emphasis—such as the make-certain program—be added or modified as corporate needs develop or change.

Although not part of his fourteen implementation steps, Crosby considers the audit a valuable tool when accomplished properly. "Let the operations conduct periodic audits of themselves, list results, and act upon them...By monitoring self-audits on a proper basis, you can cover much more ground..."

Make-Certain Program

Crosby (1979) describes the make-certain program as "a person-to-person, white-collar-oriented, improvement program that gets everyone's attention immediately." It is a training workshop for about 20 persons from administrative functions. The participants discuss their personal, on-the-job experiences in order to pinpoint weaknesses in their organizations' efficiency and to provide ideas for improvement.

The program is designed to eliminate the 25% of nonmanufacturing work that is routinely repeated. This is achieved by confirming that all activities are done correctly the first time.

It may be noted that the typical technical, clerical, or managerial job requires communication links. Thus, the effectiveness of business systems is determined by how well the paperwork, data entry, or other communication is carried out. Significantly, if a document passes through 100 operations and each operation is 99% perfect, the probability that the document is correct is only 35%.

Three specific recognitions should come from the problem list developed by the participants:

1. Paperwork and other communications systems contain most of the nonconformities in the company. Although production has its own share of nonconformities, the production cycle includes a considerable number of quality checks.

2. Every problem is preventable and the person who can best prevent it has, in the past, caused it.

3. Each individual will not believe this applies to him or her personally. That is human nature.

A program for gathering, costing, and implementing solutions to problems is then developed. The program will emphasize detail and minor changes which eliminate and prevent future nonconformities.

Management Style Evaluation

Crosby (1979) contends that "you need to take advantage of the assets you have," and, further, that knowing one's own management style will improve performance. The ten-item checklist permits a self-rating guide:

Listening   Listen and question until you understand the message. Nothing is more important—or harder to obtain—than true understanding.

Cooperating   Teamwork is learned. The whole is really greater than the sum of its parts. Dependability grows with experience.

Helping   To help in a positive manner, you must be genuinely interested in people and results.

Transmitting   There are three basic transmission modes: writing, prepared speaking, and conversation.

Creating   Problem-solving as required by managers is a learnable skill. Be careful to give appropriate credit for persons who assist.

Implementing   Real implementers are respected, appreciated, and well paid.

Learning   It is impossible to prevent things worth learning from happening.

Leading   Provide motivation for measurable, carefully defined goals.

Following   Always be able to work on someone else's dream.

Pretending   Don't. Pretending is a terrible management style.

The Crosby approach provides a look at the fundamental sources of quality costs, direction toward producing true zero defect product, and a step-by-step program for implementation.

Quality as a System

The approach that Dr. Deming uses to analyze and improve the quality level of organizations expresses closely the concepts that have become known as total quality management. In the preface to *Statistical Quality Control,* the first textbook on the subject, written by Grant and Leavenworth in 1946, is the acknowledgment of "the particular influence" of Dr. W. Edwards Deming.

Deming believes that people are a company's most important resource. These people, by working together, must solve problems and continually improve the way the company runs. Further, the cause of any current crisis or squandering of manpower is the failure of management to plan for the future and to foresee problems.

Deming directs his message toward executives, particularly CEOs and COOs. He strongly expresses the opinion that organizational change must be implemented from the top. In keeping with this philosophy, he presents improvements by measures of productivity and market position.

He also sees quality improvement starting a chain reaction (Figure 2.2). If quality is improved, then costs of rework, mistakes, and delay are reduced. This allows better use of time and materials, so productivity increases. The company can then increase market share with lower price and better quality products. The company will be able to stay in business and provide more jobs.

Deming's view of American industry is extremely negative. He is reputed to have exhorted the government to "export anything to a friendly country, except American management." Endemic to U.S. business are "deadly diseases" (Deming, 1986):

1. Lack of consistency of purpose

   A company without constancy does not see beyond the next quarter and has no long-range plans. Programs come and go.

2. Emphasis on short-term profits

   Paper profits do not make the pie bigger. Someone gets a smaller piece. Must American industry be forever subject to such plunder?

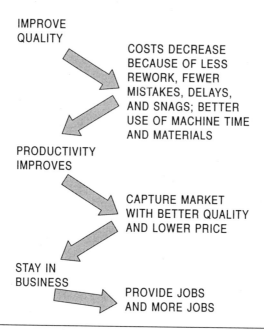

IMPROVE
QUALITY

COSTS DECREASE
BECAUSE OF LESS
REWORK, FEWER
MISTAKES, DELAYS,
AND SNAGS; BETTER
USE OF MACHINE TIME
AND MATERIALS

PRODUCTIVITY
IMPROVES

CAPTURE MARKET
WITH BETTER QUALITY
AND LOWER PRICE

STAY IN
BUSINESS

PROVIDE JOBS
AND MORE JOBS

Figure 2.2 The QI Chain Reaction (*Source:* Deming, W.E. (1986). *Out of the Crisis.* MIT CAES: Cambridge, Mass. Reproduced with permission.)

3. Evaluation of performance

This is management by fear. Measures discourage risk-taking, build fear, undermine teamwork, and pit people against each other. They encourage short-term performance, increase variation when the goal is to decrease it, and leave people bitter, despondent, dejected, and even depressed.

4. Mobility of top management

People require time to work together. Mobility creates prima donnas for quick results.

5. Running on visible figures alone

Visible figures may lead to a short-term mentality. Many important figures are unknown and unknowable. Quality costs and improvement, for example, are difficult to quantify.

Dr. Deming provides fourteen principles for transformation that must be addressed before quality can be improved. These were presented in Japan and represent his most famous contribution to quality:

1. Create constancy of purpose toward improvement of product and service, with the aim to become competitive and to stay in business, and to provide jobs.

    Rather than making money, it is management's responsibility to stay in business and provide jobs through innovation, research, constant improvement, and maintenance.

2. Adopt the new philosophy. We are in a new economic age. Western management must awaken to the challenge, learn their responsibilities, and take on leadership for a change.

    Americans are too tolerant of poor workmanship and sullen service. We need a new religion in which mistakes and negativism are unacceptable.

3. Cease dependence on inspection to achieve quality. Eliminate the need for inspection on a mass basis by building quality into the product in the first place.

    A company typically inspects a product as it comes off the line or at major stages. Defective products are either thrown out or reworked; both are unnecessarily expensive. In effect, a company is paying workers to make defects and then to correct them. Quality comes not from inspection but from improvement of the process. With instruction, workers can be enlisted in this improvement.

4. End the practice of awarding business on the basis of price tag. Instead, minimize total cost. Move toward a single supplier for any one item, on a long-term relationship of loyalty and trust.

    Frequently, seeking lowest price also leads to low quality supplies. Instead, seek the best quality and work to achieve it.

5. Improve constantly and forever the system of production and service, to improve quality and productivity, and thus constantly decrease costs.

    Improvement is not a one-time effort.

6. Institute training on the job.

    Workers often learn their jobs from another worker who was never trained properly. They are forced to follow unintelligible instructions. They can't do their jobs because no one has told them how.

7. Institute leadership. The aim of supervision should be to help people and machines and gadgets to do a better job. Supervision of management is in need of overhaul, as well as supervision of production workers.

    Leading consists of helping people do a better job and of learning by objective methods who is in need of help.

8. Drive out fear, so that everyone may work effectively for the company.

    Many employees are afraid to ask questions or to take a position, even when they do not understand what the job is or what is right or wrong. People continue to do things the wrong way, or to not do them at all.

9. Break down barriers between departments. People in research, design, sales, and production must work as a team, to foresee problems of production and in use that may be encountered with the product or service.

    Often areas are competing with each other or have goals that conflict. They do not work as a team so that they can solve or foresee problems.

10. Eliminate slogans, exhortations, and targets for the work force asking for zero defects and new levels of productivity. Such exhortations only create adversarial relationships, as the bulk of the causes of low quality and low productivity belong to the system and thus lie beyond the power of the work force.

    They never helped anyone do a good job. Let people make up their own slogans.

11a. Eliminate work standards (quotas) on the factory floor. Substitute leadership.

11b. Eliminate management by objective. Eliminate management by numbers, numerical goals. Substitute leadership.

    Quotas only account for numbers, not quality or methods.

12a. Remove barriers that rob the hourly worker of his right to pride in workmanship. The responsibility of supervisors must be changed from sheer numbers to quality.

People are eager to do a good job and distressed when they cannot.

12b. Remove barriers that rob people in management and in engineering of their right to pride in workmanship. This means, inter alia, abolishment of the annual or merit rating and of management by objective.

13. Institute a vigorous program of education and self-improvement.

Both management and the work force will have to be educated in the new methods, including teamwork and statistical techniques.

14. Put everybody in the company to work to accomplish the transformation. The transformation is everybody's job.

It will take a special top management team with a plan of action to carry out the quality mission. Workers can't do it on their own, nor can managers. A critical mass of people must understand.

The philosophy of Deming has affected many individuals, companies, and nations in the development of quality systems. As can be easily seen from reviewing the fourteen points, following Deming is not easy. The transformation he has designed is not simply the addition of another program. It is a major change in the design of the organization as well as a change in the way every manager conducts his or her affairs.

Product Design

Genichi Taguchi (1986) subtitled one of his books *Designing Quality into Products and Processes*. He has, among other achievements, taken a new look at what quality means. Rather than define quality in terms of employee activity, Taguchi concentrates on the operation of the final product. He defines product quality as follows:

Quality is the loss a product causes to society after being shipped, other than any losses caused by its intrinsic functions.

Taguchi believes that value, which usually forms the basis for quality definitions, is too subjective to measure. The parameters of value—although of vital importance to a company—are quite complex and require techniques of marketing and product planning rather than engineering. Thus, Taguchi "is opposed to treating quality questions as value questions."

There is an interplay between quality and variety, according to Taguchi. A decision as to the particular sizes within a range means that there is a loss. If, as Taguchi (1986) states, shirt neck sizes are in 1-cm increments, then a person with a 40.5 neck will have a loss equivalent to 0.5 cm due to the lack of an exact size. This problem extends to other body dimensions, colors, patterns, materials, price, and even style. However, the selection from such variety is a customer decision that is apparent at the time of purchase and will not involve the manufacturer in complaints or liability. Deciding how to segment the market and designing and pricing the varieties are problems that affect all departments, including planning, sales, and design.

There are a number of types of losses which could occur due to product use:

Loss cause	Included	Examples
1. Intrinsic function	No	Accident losses from liquor Learning losses from TV
2. Variability of function	Yes	Speed change losses in motors Color change losses in finishes
3. Harmful side effects	Yes	Noise of operating a product Vibrations caused by product

After eliminating all causes that are deliberately induced by the product and any decision about whether or not a product should be permitted in a society, Taguchi defines the modern product cycle and makes it the responsibility of quality management. The steps in the quality cycle are research, design, production, use, and disposal. Research establishes the parameters for a new product based upon customer needs. Design provides a prototype based upon the research parameters; this prototype represents the organization's highest level of quality. Production provides exact copies of the prototype; increasing variation from such exactness represents an increasing degree of nonconformity. Use is limited by the characteristics of the prototype in addition to the loss in quality due to nonconformity. Disposal is an additional cost of quality.

The Importance of Quality Design

All of the quality problems caused by variability can be eliminated or mitigated if found in design. Although product variability can be reduced in production, deterioration of the product or unsuitability for its environment cannot.

For estimating the variability of function, Taguchi uses a quadratic equation to calculate cost of the losses. The amount of cost increases as the square of the variability increases (see Figure 2.3). The function differs from the normal concept of tolerances in that items can be within specification and still have a cost for loss.

The difference between the Taguchi approach and the specification approach is represented in a case about two Sony plants (Taguchi,

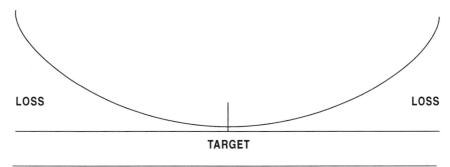

Figure 2.3 Taguchi Loss Function (*Source:* Evans, J. and W. Lindsay (1993). *The Management and Control of Quality.* West: Minneapolis, Minn. Reproduced with permission.)

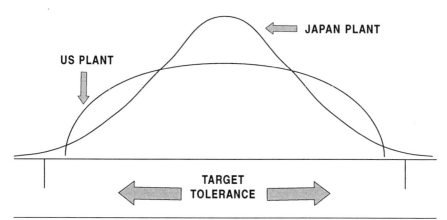

Figure 2.4 Color Density of TV Sets (*Source:* Evans, J. and W. Lindsay (1993). *The Management and Control of Quality.* West: Minneapolis, Minn. Reproduced with permission.)

1986). In San Diego, color density was always adjusted within specification. There were no defects. Workers were taught to adjust to within the tolerances. In the Japanese home plant, workers were taught to adjust to zero tolerance, that is, as close to optimum as possible. The San Diego plant had considerably higher field service costs (see Figure 2.4).

Taguchi loss function costs can be calculated for five different conditions:

1. Costs are the same on both sides.

2. Costs are different on both sides.

3. Cost reduction is smaller.

4. Cost reduction is larger.

5. Cost environment is dynamic.

Taguchi utilizes his philosophy to provide two techniques. The first is quality function deployment, a method of translating customer desires into production and quality specifications. The second is the area of design of experiments commonly labeled Taguchi techniques.

Quality Function Deployment

Mitsubishi (Evans and Lindsay, 1993) in 1979 was building ships at Kobe, Japan. Each ship was individually designed, and the builders wanted to reduce the tension between the company engineer and the customer. They realized that there was a language problem; when the customer said the product must steer easily, the engineer had to translate this into the geometry of the rudder, the controls, and the propulsion system. The customer needs, called "voice of the customer" or "customer attributes," were translated into "counterpoint characteristics" or "technical features." Thus, quality function deployment was seen as a process to identify customer needs properly at the beginning of the design process.

House of Quality

A set of four matrices is used to relate the voice of the customer to the operating instructions and quality plan. Each has a triangle interrelationship field. Some visualize it like a roof. Hence, the entire matrix looks like a house and is called the *house of quality* (see Figure 2.5).

The four matrices are labeled according to their function. Each occurs as the level of detail in the specification develops from the original customer attributes. These houses constitute major product planning documents:

1. Customer requirement planning translates the voice of the customer into counterpoint characteristics.

2. Product characteristic deployment translates product counterpoint characteristics into critical component characteristics.

3. Process plan identifies critical processes and product parameters.

4. Operating instructions identify all major operations that affect quality parameters.

The first house, involving marketing and product planning, generates the important characteristics of the product from the customer

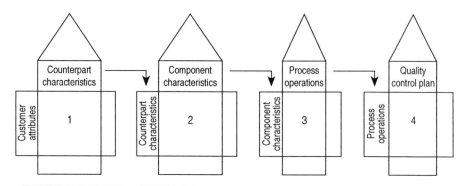

Figure 2.5 The Four Houses of Quality (*Source:* Evans, J. and W. Lindsay (1993). *The Management and Control of Quality.* West: Minneapolis, Minn., p. 163. Reproduced with permission.)

characteristics. Those considered important due to customer needs, poor competitive performance, or high marketability are deployed or carried into the second house. In the second matrix, product engineering designs subsystems and components. Each subsystem with major impact on the customer characteristics has a matrix developed. The important results of all the second house matrices are deployed into the third house. The third house matrices provide a plan for relating component characteristics to process activities. Production supervisors and operators conduct the planning and execution. The key characteristics deployed into the fourth house are the critical product parameters developed in the process. These are key control points that form the basis for a quality control plan. The fourth house matrix thus shows where and what to monitor or inspect.

Building a house of quality requires six steps (see Figure 2.6):

1. **Identify customer attributes.** Use actual customer needs, in the words of the customer. Divide the comments into primary, secondary, and tertiary categories.

2. **Identify counterpart characteristics.** Provide, in the language of the engineer, measurable technical characteristics that must be deployed throughout the design. The interrelationships are designed into the "roof" of the house of quality. Very strong, strong,

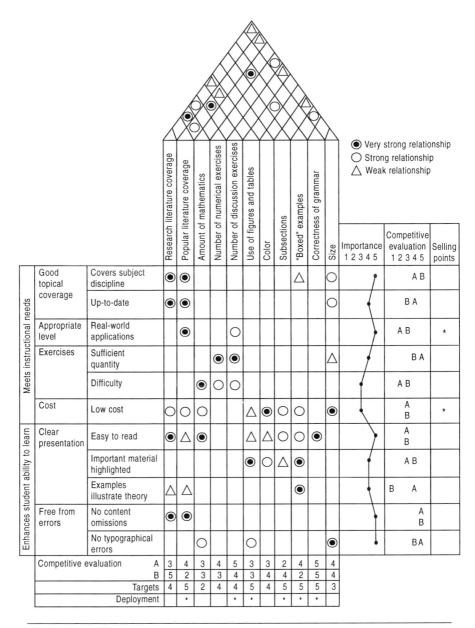

Figure 2.6 Completed House of Quality (*Source:* Evans, J. and W. Lindsay (1993). *The Management and Control of Quality.* West: Minneapolis, Minn., p. 161. Reproduced with permission.)

and weak relationships are identified. Strong relationships show features that must be approached collectively rather than individually.

3. **Relate the customer attributes to the counterpoint characteristics.** Develop, in the body of the chart, the relationships. Lack of any strong relationship for a customer attribute with a counterpart shows that attribute has not been addressed and the product may not be successful. If the counterpoint does not relate to a consumer characteristic, it may be redundant or designers may have missed a consumer attribute.

4. **Conduct an evaluation of competing products.** If a competitor's product receives a low evaluation for an attribute, it may be a focus for competitive advantage. Selling points and evaluation of importance are added.

5. **Evaluate counterpoint characteristics of competition and develop targets.** On the basis of technical evaluations, develop targets based upon opportunities in competing product characteristic.

6. **Determine which characteristics to deploy.** Select those critical factors to be deployed. Care must be taken to ensure that the same attitude expressed by the customer is incorporated throughout the process.

The vast majority of applications in the United States involve the marketing and design aspects, the first and second houses. Because numerous authors have cited the lack of cross-functional project successes, more significant benefits may be available in the third and fourth houses of quality.

Design of Experiments

Taguchi and his followers—who include many world-class companies—have developed extensive case histories and applications for a technique to analyze product and process. Three levels of application are cited: system design, parameter design, and tolerance design.

System Design

Initial setting of parameters is established through evaluating customer needs and manufacturing limitations. Utilizing the best available technology, a basic design is developed. Functional, technological, and economic influences must be taken into account.

Parameter Design

Products insensitive to the range of variations in the environment are, according to Dr. Taguchi, robust. The high-quality product will have such robustness in many of the most important features. By utilizing signal-to-noise ratio, a measure common to electronic communications, he is able to compare the relative robustness of product attribute combinations.

Experimental design techniques were developed by Fisher in the 1920s but not widely used. The large number of factors and interactions caused data collection and data interpretation problems. Taguchi utilized a noise factor to contain many of the minor variables, permitting the reduction in variables from hundreds to units. Often, the insurmountable problem of 165 variables becomes the undergraduate symbol of triteness, the three-variable problem.

Taguchi's parameter design experiments attempt to reduce variability, which is useful to both the design and quality process engineer. They do, however, violate some traditional principles and have been criticized by the statistical community. Although certain experimental design issues are subject to debate, Dr. Taguchi must be recognized not only for popularizing experimental design, but also for pioneering the simultaneous study of mean and variance, popularizing robust design, and focusing attention on the costs associated with product quality.

Quality Culture

Joe Juran (1987) views the process of improving quality more as evolutionary rather than revolutionary. He sees that there are various cultures, including rules, mores, and languages, in the corporation. It is through communicating in the appropriate language that quality improvement best can be installed. Where upper management is addressed in terms of cost accounting and other dollar comparisons, operational personnel are addressed in terms of elimination of defects and statistical control techniques. Juran views the pursuit of quality on two levels: (1) the mission of the company as a whole is to achieve high quality and (2) the mission of each unit is to deliver high production quality.

Quality as defined by Juran is "fitness for use." This fitness can be measured by four parameters: design, conformance, availability, and field service. Design involves the market research and subsequent specifications. Conformance is manufacturing to the design and subsequent customer feedback. Availability is the reliability and maintainability of the product. Field service includes integrity, promptness, and competence.

Juran (1987) sees quality as three processes, which he calls the quality trilogy: quality planning, quality control, and quality improvement. It is through completing activities continuously in each of these areas that continuous quality is attained.

Quality planning is the process of preparing appropriate quality goals. This includes a standardized series of activities:

1. Identify customers

2. Determine customer needs

3. Develop product features to meet needs

4. Establish product goals

5. Develop processes to meet product goals

6. Prove process capacity

Quality control is the meeting of quality goals and reduction of variation by the operating forces. It includes location and elimination of significant chance variables, the latter which Juran labels "sporadic spikes."

Quality improvement (Juran, 1987) results from outperforming the past. Here, stable control is established after a significant improvement in quality. The chronic level of deficiencies has been modified. The process consists of the following steps:

1. Identify specific projects for improvement

2. Organize project teams

3. Discover causes

4. Develop remedies

5. Prove effectiveness of remedies

6. Deal with cultural resistance

7. Establish controls to hold gains

In implementing quality improvement in a company, Juran expects to find an active quality control organization with very little planning or improvement. His approach is, then, to develop management planning that will support and implement the results of the individual project teams.

Juran has developed a series of 24 benchmarks for executives to assess the quality program:

Quality Vision

1. Quality vision	A general statement of what the organization does and its planned level of performance

Strategy and Deployment

2. Strategy with customer focus	A quality strategy is based on the vision, customer needs, and the customer view of quality
3. Quality goals	Specific, measurable, and related to product quality
4. Deployed goals	Group and individual assignments and existing or cross-functional teams
5. Monitoring progress	Results for each goal

Organization and Resources

6. Quality council	Upper management steering committee
7. Set priorities	Project-by-project improvement

8. Establish teams	Formal, mandated, structured process, important cross-unit projects, and product development projects
9. Provide resources	Facilitator support, training, and time for meetings and analysis
10. Review progress	Team reports about if the team is following methodology, roadblocks needing upper management, data limitations, and plan for next period

Data and Information

11. Customer data	Identification of customers: customer needs expressed in terms of benefits, customer needs as priorities, customer evaluation of competition, and measures of competition's product performance
12. Costs of poor quality	Costs of inspection: costs of rework or repeat work and costs of customer service
13. Internal culture information	Organizational strengths and weaknesses and improvement opportunities addressing employee needs
14. Audit of quality systems	Upper managers develop key questions and subordinates gather data

Quality Culture

15. Leadership style	Upper management style of actions
16. Personal involvement	Seminar attendance, serve on project teams, meet with customers and employees, and include quality routinely on agendas
17. Consistency	Not allowing other events to slow quality progress
18. Elimination of blame	Blame prevents identifying improvement opportunities
19. Data-based decisions	Quality depends on good data

20. Changing culture through action — Attitudes change as a result of participation and type of training is an indicator of effort's direction

21. Changing culture through participation — Participation helps reduce resistance

22. Changing culture by providing time — People often require time to adjust to changes

23. Motivation through recognition — Recognition is a strong nonmonetary incentive to change, but it must be in a form that is valued by honored individuals

24. Motivation through financial reward — Quality is part of the job and should be one of the primary determinants in performance evaluation

Juran has developed an approach that considers the cultural environment of the enterprise. As a result, resistance to change is taken into account during implementation. Although the line worker is thus considered, Juran focuses attention on top management. He expects traditional lines of authority, bolstered by recognition and audit teams, to revolutionize quality attitudes.

TOTAL QUALITY CONTROL: THE ISHIKAWA APPROACH

"My wish is to see the Japanese economy become well established through QC and TQC and through Japan's ability to export good and inexpensive products worldwide. It will then follow that the Japanese economy will be placed on a firmer foundation, Japan's technology will become well-established, and Japan will be in a position to engage in the export of technology on a continuous basis. As for companies, I hope they can share their profits with customers, employees, shareholders, and society in general. I hope these companies can become instruments for enhancing the quality of life not only of the Japanese peoples but also of all peoples, and in this way bring about peace in the world." (Ishikawa, 1985)

K. Ishikawa is perhaps the most influential and outspoken of the new wave of Japanese quality leaders. His involvement began during the post-war period, and he rose to become the definer of what total quality management (TQM) means to modern Japan. His vision of the role of the individual, company, nation, and world may ultimately make him the premier quality philosopher. At the very least, he provides many clues as to the difference between the American and the Japanese approaches.

The Japan that Ishikawa describes is not the one commonly assumed by our stay-at-home American industry experts. Although it was severely damaged by World War II, Japan had a sophisticated infrastructure. Compulsory education and public schools had existed since the Meiji Restoration (1868). Art and architecture, both of Japan and of the world, flourished. In addition to its unique classical pottery styles, industrial Japan was building structures designed by such eminent architects as Sullivan, Wright, and van der Rohe. Automobiles and appliances were made on modern assembly lines, and as early as 1928 the automobile outnumbered the rickshaw. Japan soundly defeated tsarist Russia in 1898 with a modern fleet, participated in World War I against the Imperial German Fleet, and entered World War II with a strong superiority in naval and aviation equipment. Prior to World War II (1935), Japan adopted a version of British Standards 600, which would later become U.S. Z-1. The task facing a devastated post-war Japanese industry was one of recreating what had already existed, rather than creating an industrial base from a rice-patty mentality.

It is not surprising that Ishikawa felt that the Japanese quality revolution may not be duplicated outside Japan. Post-war Japan was uniquely qualified with persons of skill and will for the task at hand. It is interesting that Ishikawa, as an instructor in the Union of Japanese Scientists and Engineers (JUSE), became familiar with both Deming and Juran.

What Is Quality Control?

Ishikawa (1985) has a unique definition of quality control as a practice or activity:

> To practice quality control is to develop, design, produce, and service a quality product which is most economical, most useful, and always satisfactory to the customer.
>
> To meet this goal, everyone in the company must participate in and promote quality control, including top executives, all divisions within the company, and all employees.

Some points to remember about implementing quality control are the following:

1. Companies engage in quality control in order to satisfy the requirements of the customer.
2. Companies must emphasize customer orientation.
3. Quality includes product and service.
4. Quality includes a fair quality in a fair amount for a fair price.

The Quality Thought Revolution

Ishikawa proposes a six-concept structure as a remedy to what he sees as Japanese management acting irrationally in not meeting the definition of product quality. Utilizing these six concepts, organizational quality can be improved:

1. Emphasize quality first, not short-term profit.
2. Focus on customer orientation, tastes, and needs.
3. Consider the next process in the production cycle as a customer.
4. Use data and facts for presentation and statistical methods.
5. Incorporate respect for humanity into management philosophy.
6. Employ cross-function management.

Ishikawa is an engineer and a manager, as evidenced by his comment to "...group psychologists who want a piece of the action. There are theorists who create theory X, theory Y and theory Z and provide their critiques of our activities....All of such theories are contained in our Quality Circle activities. We do not present them as theories, however, we simply practice them."

Knowing True Quality

The traditional way of evaluating quality is via conformance to specifications. Ishikawa states that "product standards and tolerances limits on a drawing are quite unreliable." He warns that product standards, material standards, tolerance limits, and data from physical or chemical measures are to be treated "with skepticism." None of these measures are necessarily true measures of product quality. Even if such measures were requested by the customer, they may not be true measures.

True measures are expressed in terms of customer need and in ways the customer can understand. Such terms may be difficult to translate into a measurable characteristic. This is one task for which the cause-and-effect, or Ishikawa, diagram (the name coined by Dr. Juran in the 1962 QC Handbook) is useful (see Figure 2.7).

For example, a newspaper company wants paper that won't tear on the press. However, measuring this characteristic would require actually using it on the press, an impractical test. Therefore, measures such as width, thickness, tensile strength, color, and rolling performance must suffice. These substitute characteristics can then be measured. The relationship between the true and the substitute characteristics can then be determined.

The actual procedure for implementing the process of determining the substitute measurement characteristics is:

1. Understand the true characteristic
2. Determine how to measure the true characteristic
3. Discover substitute characteristics and relations

The process is described by Ishikawa as being quality function deployment. It is described in this chapter under the heading of Taguchi because of subsequent work by the American Supplier Institute in formalizing the procedure.

Expressing the Quality

In order to express quality, certain common parameters must be established. Ishikawa identifies seven:

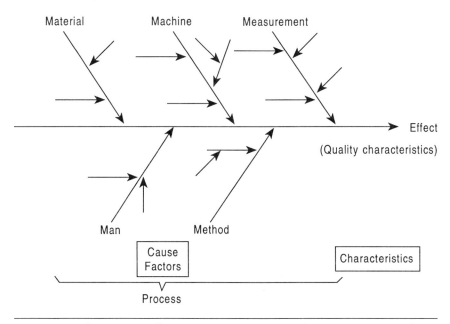

Figure 2.7 Cause-and-Effect Diagram (*Source:* Ishikawa, K. (1985). *What Is Total Quality Control? The Japanese Way.* Prentice Hall: Englewood Cliffs, N.J., p. 63. Reproduced with permission.)

1. The assurance unit (population or lot size)

2. The measuring method (perception or test)

3. Relative importance of characteristic (not equally important)

4. Consensus on defects and flaws (actually unusable)

5. Expose latent defects (go-straight-percentage)

6. Observe quality statistically (distributions)

7. Quality of design (improvement in product)

The Quality Circle

The quality circle movement, chronicled and publicized by Ishikawa in his journal, *QC for Foremen,* includes 1.5 million persons registered with the Japanese QC Circle Headquarters and an estimated ten times that number who are not registered. According to Ishikawa, perhaps 17 million workers in all are involved in circles.

Quality circles are a part of company-wide quality control (QC) activity. The foremen and workers must assume responsibility for the process in order for QC to become successful. Such QC activities mirror the abilities of the management leadership (good leaders will have active circles, and poor leaders will have poor circles). The actual activities of the circles are consistent with human nature and, thus, could be successful anywhere in the world. Where there are no QC circle activities, according to Ishikawa, total quality control activities cannot exist. There are three basic goals for the circles:

1. Contribute to the improvement of the enterprise

2. Respect humanity and build a meaningful work environment

3. Exercise human capabilities fully and draw out individual potential

Quality circles, as designed by Ishikawa and his associates, provide five major functions to achieve stability and achievement of the goals:

1. Volunteerism	Democratic management
2. Self-development	Individual study
3. Mutual development	Broader perspectives
4. Participation by all members	Everyone completes tasks
5. Continuity	Not same as taskforce

Quality Circle Activities

In reviewing the results of Japanese QC circles, Ishikawa found that they contributed to the company in three ways:

1. Contribute to the improvement and development of the enterprise

2. Respect for humanity and build a worthwhile, happy, and bright workshop

3. Exercise human capabilities fully, and eventually draw out infinite possibilities

Quality Circle Evaluation

The Japanese evaluate quality circles over a much wider criteria than monetary results. Reported results have many factors, and transferability of such is questionable. In evaluating projects of QC circles, Ishikawa provides the following criteria:

1. Theme selection	20 points
2. Cooperative effort	20 points
3. Understanding problem and results	30 points
4. Results	10 points
5. Standardization and solution effect	10 points
6. Reflection (recommendations)	10 points

Why Did Zero Defects Fail?

In 1965, Ishikawa observed the Zero Defects (ZD) movement and found the activities to be different from the already successful Japanese quality circles in terms of a number of key points. He was troubled that the American process seemed to be temporary and potentially doomed. At that point he described what he perceived as shortcomings in order to prevent repetition of errors:

1. The system emphasized individual will; each person does his best.

2. It was a movement without tools and with no process to change methods.

3. Standards correction was left out. Good product was expected to grow from adherence to standards.

4. Workers were expected to follow orders more exactly.

5. Kickoff really was an attempt to provide some enthusiasm.

6. All responsibilities for mistakes were the employee's.

7. The movement became a big show.

8. There was no mutual development group, national center, or organization.

Financial, Performance, and Quality Audits: A Comparison of Similar Approaches

Financial auditing is a long-standing practice to determine the fiscal health of an organization. It is a review of the past, a sampling of the present, and a prediction of the future. The financial audit conclusion is often included with the annual report, prospectus, or other important documents. In addition to the obvious reporting and prediction of profit and distributed equity, such information as the source of costs and revenues can be determined from it. Just as the financial audit is the basis for corporate decisions about how the money resource is to be collected and utilized, the quality audit is the basis for decisions about the quality process, program, and activity.

Auditing evolved from the simple questions about the financial status of an organization because questions about the potential future earnings were found to be dependent on the effectiveness of decision-making. Auditing methods, called performance auditing, were designed to test the effectiveness of the process. Then, as the importance of quality performance in the context of the entire firm became apparent, a specialized type of audit for quality systems was developed. Such quality audits have been initiated by each of the quality pioneers and are considered an important part of any quality system.

The tools of auditing have developed over the years. Financial auditors, faced with the problems of reliability and accuracy, developed standardized methods, criteria questions, and other working papers. For quality audits, working papers provide the following:

1. Memory prompters to ensure that no element is forgotten

2. A scale of effectiveness for each of the elements

3. A historical record of audit procedures, audit methods, and audit findings

The philosophy of improvement so essential to all modern quality programs is also the essential element of the quality auditing process. Each individual tool or technique will, by the nature of change, become obsolete at about the time it is implemented and working. Auditing is thus a process in which the tools of the auditor and the tools required by the auditee are expected to improve over time. What was common practice at one audit date will be not acceptable at some time in the future. Also, as quality theory becomes more accepted, an organization will need to show compliance to accepted quality practices.

The audit should be welcomed by the auditee as a means of gaining knowledge and improving operating procedures. All audits are a learning experience for each participant. As such, they should be a positive reinforcement for a job well done. It is the responsibility of the auditor to ensure that the appropriate atmosphere and decorum are maintained so that the auditee is sufficiently praised for those things being done correctly and is, therefore, motivated to improve those areas of nonconformance.

ISO 9000 AND MANAGEMENT BY OBJECTIVES

One basic tenet of total quality management programs requires that the company establish goals for the continuous improvement of quality and as a focus for the organization. Such goals require the active involvement of a senior operating officer, often the company CEO. Such leadership is considered so important that it is a nondelegatable task. None of the major quality leaders would work with a company that delegated such responsibility.

Another tenet requires that each team or circle clearly define the objective of each project and estimate the impact of such activity prior to implementation. Projects are not considered complete until such measures are verified or rejected. The results of such improvement are the true measure of team effectiveness.

Quality auditing, by its nature, tends to measure conformance of individual and group activity or results to the stated procedures of a company or accepted practices. Responsibility for procedure development and all subsequent implementation activities is the responsibility of various members of the management. Often, there is documentation required by the audit that specifies the responsible position.

Thus, a formal quality goals structure with assigned responsibility would appear to be mandated in the name of total quality management, team building, and quality auditing. If this were truly a totally rational world, such goals would be required.

This is, however, not a rational world. Over the past hundred years, movements to improve industry have been instituted only to deteriorate into processes to eliminate nonproductive managers. Motivated by fear, managers would respond by forecasting improvements in inconsequential areas. In meeting these forecasts, improvements and subsequent managerial promotions occurred to the overall detriment of the company. Included in the fourteen points of Deming, the cost calculations of Crosby, the managerial caveats of Juran, the quality function deployment process of Taguchi, and the descriptions of recent Japanese quality movement by Ishikawa are implicit and explicit warnings against using fear as a personnel management tool and greed as the organizational goal.

The authors anticipate that management goals and tracking progress toward those goals will take on greater importance in future editions of ISO 9000.

ENDNOTES

Crosby, P. (1979). *Quality Is Free.* Mentor: New York.

Deming, W. E. (1986). *Out of the Crisis.* MIT CAES: Cambridge, Mass.

Evans, J. and W. Lindsay (1993). *The Management and Control of Quality.* West: Minneapolis, Minn.

Ishikawa, K. (1985). *What Is Total Quality Control? The Japanese Way.* Prentice Hall: Englewood Cliffs, N.J.

Juran, J. (1987). *On Quality Leadership.* Juran Institute: Wilton, Conn.

Taguchi, G. (1986). *Introduction to Quality Engineering.* Asian Productivity Organization, American Supplier Institute: Dearborn, Mich.

CHAPTER 3

APPLYING THE QUALITY SCIENCES TO ISO 9000

HOW DOES QUALITY APPLY?

If most business managers were asked how quality applied to their businesses, they would most likely expound upon the virtue of a service which satisfies the customer. This is because most literature written on quality focuses on the service or process to produce the service. Actually, before discussing quality in business there must exist some business model to which it applies. Describing quality before defining the business is akin to having an answer and finding the facts to fit it.

Defining "business" may seem to be an easy matter, but there are literally hundreds of business models from which to choose. One of the better definitions by organizational theorists states that businesses are made up of groups of people striving to accomplish and acquire things that they could not accomplish or acquire individually. Businesses do not exist just to make a profit. Most studies performed today indicate that people rarely get into business primarily for profit. Rather, most people start a business as a means of independence and self-expression. The definition includes all of the elements of a general systems feedback loop. People do things (transformation) that result in an accomplishment (output) and acquisition (input) of things that could not be done individually (feedback evaluator). See Table 3.1 for an example of a business model.

Table 3.1   Business Model

Organizational model	A business system model given a level of abstraction for open systems that reflects the basic systems concepts and characteristics, but which is more usable as an analytic tool.
Inputs	Those factors that are, at any point in time, givens that face the organization. These factors include environment, resources, history, and strategy.
Transformation	Those components within the organization that enable management, given an environment, a set of resources, and history, to take a strategy and implement it to produce effective organizational, group/ unit, and individual performance. These components include the activity, the individuals, the formal organizational arrangements, and the informal organization.
Outputs	What the organization produces, how it performs, or how effective it is globally. This includes goal attainment, resource utilization, and adaptability.

As can be seen from the review of the philosophies and techniques of the quality experts, defining quality is at least as difficult. Most people have their own definition. There are those who see quality as the value of an item. Others see quality as perfection. Still others see quality as a defect-free service. The first activity in establishing an ISO

9000-registerable quality system will be to establish a workable definition. This concept of quality will be expressed as a mission statement, that is, the company's vision and commitment to those who interact with it. Key concepts to be considered in defining quality in an organization can be developed by asking how a company:

- Provides customers with products and services that address their needs (sometimes unrealized or unarticulated)
- Strives to meet or exceed customer expectation (customer satisfaction)
- Fosters an atmosphere of continuous improvement (company culture)
- Develops respect and cooperation with suppliers and employees

QUALITY IN BUSINESS

When it comes to business and its operation, quality must be viewed in terms of service performance by the company in response to customer needs (see Table 3.2). Service performance includes service, product, and all the factors that must accompany delivery. Just as quality has been shown to have a number of definitions, types of performance measurement can vary greatly among business profes-

Table 3.2 Quality Definitions

Quality system	The efficiency of a system to meet external needs
Efficiency	The degree to which the external needs and demands are met by the goals, objectives, and/or structures of the system
System	A set of interrelated elements that takes input from the environment, subjects that input to some form of transformation process, and produces output
External needs	The requirements of the environment that must be satisfied in order to acquire benefits that cannot be achieved internally

sionals. The most common norm for measuring business performance is through financial means. In short, the difference between profit and loss reflects the rate of success. There is, however, a difference of opinion as to whether or not financial measures truly indicate the actual performance of a company.

Financial performance indicators only report on the outcome of business operations. A business may remain solvent for a period of time before declines in revenue begin to affect its survival. If a business only generated $50,000 in sales for the first two months of a new year while its yearly operating expenses were $25,000, it would show a net gain of $25,000, even though no sales were made the rest of the year. Obviously, in this case, financial measures of performance are not adequate, because it is assumed that the measure reflects a continuous flow of sales and expenses. Financial performance indicators may not always not show how effectively business operations are managed.

Over the past 20 years, chief executive officers (CEOs) have discovered that financial indicators are poor measures of corporate performance. Financial reports can only identify that a problem exists. Poor earnings are a sign of serious root-cause problems within a company. The question still remains, "How to accurately measure corporate performance?" Even more importantly, "How to turn a company around once the negative economic symptoms become apparent?" There has been a proliferation of quick-fix ideas to turn companies around. The most notable ones are those which result in "management bashing," replacing mid-level managers who appear to be poor performers.

The problem CEOs have with using the skills of the quality professionals is due to a conflicting view of their role as problem solvers. For the most part, CEOs view the quality organization as a means of control. In fact, many business texts include the quality organization in the control function of business operations management (implying that the quality organization exists to prevent bad products from being found by the customer in unreasonable amounts). Obviously, this gives most managers a distorted understanding of the quality organization's objectives. Couple this with a product-oriented definition of quality (the quality organization looks for bad things), which further dilutes the quality organization's true function, and the result is an ineffective group.

The main purpose of a business is to provide products or services to satisfy some need. This need is external to the business. Even a (vendor) department that services another (customer) department satisfies such an external need. To operate, a business cannot service itself; it must service an external customer or need (output) in exchange for resources (input). Quality, in this sense, is the performance of the business to satisfy those external needs.

Quality is, therefore, "the ability of a business system to utilize resources to meet external needs." The efficiency of an organization to provide such quality is the primary measure of performance.

The primary role, then, of the Quality Department is to identify, analyze, summarize, and report the efficiency of the business system in meeting external needs. In addition, the quality group may be called upon to provide performance planning, performance improvement, and performance control of the business system. The Quality Department, or at least that portion satisfying the above activities, can be called the Performance Research and Development Department.

QUALITY ORGANIZATION

Ultimately, the senior management staff is responsible for the performance of their business system, in much the same way that they are responsible for the management of company financial resources.

The key to business system performance lies in the level at which the various departments responsible for service development interact together. The service-developing departments (mainly marketing, service development, and service operations) must interact effectively in order to efficiently meet external needs. Additionally, these groups must have optimum internal tasking with checks and balances. This means that each service-developing department must continually seek ways to improve its internal performance, as well as interface with each other department (Table 3.3).

The organization of a company can have dramatic effects on business system performance. If departments are organized in such a way that their objectives conflict, an informal organization will emerge that will have hidden objectives that counteract company goals.

Table 3.3 Quality Management

Mission	Identify, analyze, summarize, and report the efficiency of the business system. Manage and develop the means for performance improvement.
Performance planning	Identification of environmental elements outside the boundaries of the organization. Development of a strategy to utilize/configure the business system to satisfy the elements. Define a feedback system to verify the strategy performance.
Performance improvement	Identification of low-performance areas within the system. Formulation of a strategy to effect performance improvement. Implementation of the performance improvement action plans.
Performance control	Development of performance standards, measurement variables, and measurement methods. Measuring performance and reporting departures from standards. System intervention to maintain standards.

When this happens, business efficiency for meeting external needs will decline. This decline leads to a drop in revenue. As the efficiency for meeting external needs declines, there is an associated decrease in sales that occurs within six months to two years.

The lag between decline in efficiency and sales is due to efforts by management to reduce direct costs (labor and material) and ineffective market promotional efforts. When management attempts to counteract poor efficiency through reduction of direct costs, the result may be an acceleration of the decline, rather than efficiency improvement. This is due, in large part, to the fact that performance planning and control have not been improved.

It is important to note that regardless of the service a business provides, three basic groups of employees are prevalent in any business (see Table 3.4). These groups are decision-makers, service developers and internal analyzers. Each has specific objectives that differ significantly from the others. Any mixing of these groups, such as placing the Accounting Department under Service Operations, will cause a decrease in business efficiency. In the same manner, placing

Table 3.4 Business Elements

Organizational system elements	Organizational systems are composed of internal interdependent components or subparts. The organization also has the capacity for feedback, equilibrium, and optimization.
Decision-makers	Managers responsible for resource allocation and configuration to implement a strategy and produce effective organizational, group/unit, and individual performance.
Service developers	Those groups/units within the organization that develop, transform, or provide output.
Internal analyzers	Those groups/units within the organization that provide feedback on the performance of the organizational system.

the Marketing Department under Human Resources would cause the same conflict in objectives, leading to decreased business efficiency. Accordingly, placing the Quality Department under Service Operations will have the same effect. The relationship among these three groups is shown in Figure 3.1.

BUSINESS MODELING

Any business model used to define quality must be quantitative for analysis work. The three core elements—decision-makers, service developers, and internal analyzers—are affected by the influences of input and transformation structure. Each group/subpart in a business has a specific task orientation. For example, each of the product development groups—marketing, service development, and service operations—is involved in generating a service or product. The accounting, quality, and human resources groups monitor the performance of the business, which is a service to other groups. Finally, there are the decision-makers, the individuals who are responsible for the intergroup configuration and resource allocation to produce effective organizational performance. How efficiently these three elements operate

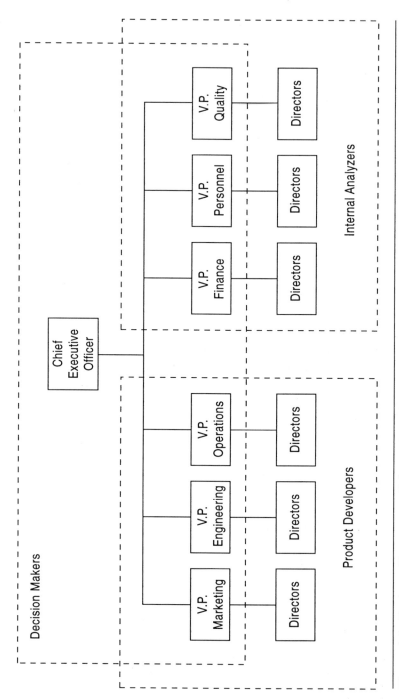

Figure 3.1 Business Department Relationships (*Source:* Mauch, P. (1993). *Basic Approach to Quality Control and SPC.* ASQC Quality Press: Milwaukee, p. 3. Reproduced with permission.)

together is influenced by four factors: inputs, transformation, outputs, and performance.

Inputs

Input to the business (see Table 3.5) and its components is comprised of four elements: environment, resources, history, and strategy. The environment may include factors over which the business has little or no control, such as the customers' needs and wants. Marketing groups can create desire, not needs. Needs are inherent to the customer.

Table 3.5 Input Elements

Environment	All factors, such as institutions, groups, individuals, and events, outside of the boundaries of the organization being analyzed, but having a potential impact on that organization
Resources	Various assets that the organization has access to, including human resources, technology, capital, and information, as well as less tangible resources (such as recognition in the market)
History	The patterns of the past behavior, activity, and effectiveness of the organization that may have an effect on current organizational functioning
Strategy	The stream of decisions made about how organizational resources will be configured against the demands, constraints, and opportunities, within the context of history

Resources used by the business and its components are limited, but their scope includes more than material. Market recognition, good will, and human resources are also included.

The past patterns (history) of organizational behavior have an effect on the performance of the business system. Of course, environment, resources, and history have dramatic impacts on the strategies the decision-makers develop.

Transformation

The transformation process (see Table 3.6) relates to both intra-group and intergroup goal achievement methods. There are four elements associated with the transformation process: task, individual, formal arrangements, and informal arrangements.

Table 3.6 Transformation Elements

Task	The basic and inherent work to be done by the organization and its parts
Individual	The characteristics of individuals in the organization
Formal organizational arrangements	The various structures, processes, and methods that are formally created to get individuals to perform tasks
Informal organizational arrangements	The emerging arrangements, including structures, processes, and relationships

Traditional quality studies have explored the task function of the transformation process in great detail. However, more coordination work needs to be done to educate the quality professional about studies done in the area of formal and informal work arrangements. Behavioral scientists have conducted studies of such work structures in business management systems for well over 80 years.

Output

Output from the business (see Table 3.7) and its components is comprised of four elements: product, customer service, societal impact, and performance records. The product may consist of goods and

Table 3.7 Output Elements

Product	All primary products and services delivered to a customer in the outside environment, including other units in a company.
Customer service	Various services provided that enhance the customer's appreciation and use of the primary product or service.
Societal impact	The effect of the product on others in the society.
Performance records	The company documents, internal and external, relating to the series of activities that led to the delivery of the product. Such documents would be held for evaluation against the demands, constraints, and opportunities within the context of product history.

services. In addition, delivery time, storage, delivery, training, and other activities are included in customer service.

The societal issues include points such as disposal, fuel use, effect on future generations, noise, safety, and health. Performance records are kept for evaluating how well the product meets the customers' needs and expectations.

QUALITY ORGANIZATION AS THE FEEDBACK MEASURE

Repeating the discussion of the role of the Quality Department under the systems model (Table 3.3), its activities are identifying, analyzing, summarizing, and reporting the performance of the business system. It also requires the quality organization to perform research and development of business performance. An example of this would be evaluation of proposals to determine the performance of the service development group. Quality audits are also a means of determining the level of performance of a system.

Because of the stereotypical picture associated with the term "quality control/assurance department," a better reflection of the organization's role would be "performance research and development" (see Table 3.3). This would present management with a clear understanding of the organization's objectives. A parallel to this would be the Accounting Department. The accounting group monitors, reports, and manages the financial performance of the business. Conversely, the performance research and development group monitors, reports, and manages the efficiency of the business system.

The appropriate ISO 9001 paragraphs that apply to the feedback model are listed according to their relationship to the three activity groups and their function in the system in Table 3.8.

DEFINING THE SERVICE

Since quality is measured by the efficiency of a system to meet external needs, the needs must described in some manner. In the ISO 9000 standards, 4.3 Contract Review is used to verify that external needs are met. However, that section is only one point where service delivery is important. Throughout the organization, tasks need to be focused toward the efficiency of delivery of quality services meeting customer needs. There are various means other than contract review for evaluating customer wants and needs, and some include quality function deployment and benchmarking. A useful framework from quality assurance techniques is to approach service in terms of risk analysis. Regardless of the means used to define external needs, the procedure must be identified, established, and maintained in ISO 9000, Section 4.20 Statistical Techniques and appropriate records of analysis results maintained in order to evaluate how well the business systems have met them.

The first essential step in planning control and improvements of methods used in the business is defining the customer requirements in terms of the service provided. Describing the service will help eliminate waste and give a more focused approach to those who do not have direct contact with the customer. It will also help to determine the correct actions to take when changes are needed.

Table 3.8 ISO 9000 Activity Group Responsibility

	Priority	Imple-mentation	Oper-ation	Audit
4.1 Management Responsibility	1	DM	DM	DM
4.2 Quality System	1	DM	CS	CS
4.3 Contract Review	3	PD	PD	CS
4.4 Design Control	4	PD	PD	CS
4.5 Document and Data Control	2	CS	CS	CS
4.6 Purchasing	3	PD	PD	CS
4.7 Control of Customer-Supplied Product	7	PD	PD	CS
4.8 Product Identification and Trackability	6	CS	PD	CS
4.9 Process Control	5	PD	PD	CS
4.10 Inspection and Testing	6	PD	PD	CS
4.11 Control of Inspection, Measuring, and Testing Equipment	6	CS	CS	CS
4.12 Inspection and Test Status	6	CS	PD	CS
4.13 Control of Non-Conforming Product	6	CS	CS	DM
4.14 Corrective and Preventive Action	2	PD	PD	DM
4.15 Handling, Storage, Packaging, and Delivery	7	PD	PD	CS
4.16 Control of Quality Records	7	CS	CS	CS
4.17 Internal Quality Audits	2	CS	CS	CS
4.18 Training	6	PD	PD	DM
4.19 Servicing	7	CS	CS	CS
4.20 Statistical Techniques	4	CS	CS	DM

Note: DM = decision-makers, CS = customer service, PD = product developers. Priority is based on implementation schemes.

Service Analysis: Telemarketing Example

Telemarketing is used as an example to illustrate how a service is described. Some of the pitfalls of insufficient documentation are described and one means of documenting the telemarketing service is explained. It will be referred to later to explain how to utilize more advanced analyses. One can then compare the telemarketing services example to the definition and evaluate the difference.

Dilemma of Inexact Definitions

In many cases companies do not define their services. This causes confusion because everyone in the company will have a different idea of what the service should be. When this happens, more time is spent arguing about what the service should be and, consequently, problems seldom get solved. To avoid this dilemma, services should be defined in exact terms.

Risk Evaluation Criteria

Another aspect of defining a service involves risk. Each characteristic or variable has an associated risk, relative to the terms of how important such service is in meeting customer needs. Each service characteristic must be given a risk level assessing how it will affect these needs. Three commonly used risk levels are:

Critical risk   Any case where a deviation would cause a safety or health hazard or prevent the service from performing its basic function

Major risk   Any case where a deviation would cause a substantial reduction in the service's ability to operate within the expected operating environment or is readily noticeable by the customer

Minor risk   Any case where a deviation would not cause a reduction in the service's operating conditions and is not readily noticeable by the customer

Characteristic Parameters

A service can be evaluated with characteristics from one of three parameters: attributes (defined as conforming or nonconforming), variables (defined as a value in a range), and operating (or environmental) conditions (defined as interacting but external to the problem). The following examples illustrate how these parameters are used.

Attributes

Attributes are defined as service characteristics or features. For example, a car wash either waxes a car or it does not. A radio either plays in stereo or it does not. In the telemarketing example, the telemarketing either obtains sales or it does not. A list of attributes for telemarketing is shown in Table 3.9.

Table 3.9 Service Attributes for Telemarketing

Item	Characteristic	CR	MA	MI
1	Pleasant voice	XX		
2	Memorized script	XX		
3	Cleanness			XX

The service attributes are listed in their order of importance in Table 3.9. The columns on the right side of the table indicate critical (CR), major (MA), and minor (MI) risks. For each service feature, risk factors are assigned with regard to the five basic human needs.

As seen in Table 3.9, there are no numbers assigned to the features. The service (telemarketing) either has these features or it does not. This is what separates attributes from other descriptions that are used. Attributes can be found listed on some consumer products. A service is purchased because of the features that it has. In fact, one way to list the attributes of a service that is already being produced is to look at its sales literature. Most marketing professionals define services by features.

Table 3.1 can be defined as a service specification. Most quality professionals develop such specifications to define the service. A list of service features will usually be more detailed than those given in Table 3.9.

Attribute specifications can be found in most inspection departments. They are used to visually inspect and test services. There are many functional tests that are performed to ensure that service features operate as they are supposed to.

Variables

Variables are defined as "service characteristics containing measurable values within a potential range." Examples of variables are the mileage reading on a car's odometer or the amount of milk poured into a measuring cup. A list of service variables for telemarketing operations is shown in Table 3.10. These are characteristics that can be measured using test equipment.

Table 3.10 Service Variables for Telemarketing

Item	Parameter	Nominal/unit	CR	MA	MI
1	Call introduction	2 min.	±0.25	0.10	0.05
2	Call length	12 min.	±0.25	0.10	0.05

The number of variables can be as many as needed. As with attributes, each number must be ranked according to the level of risk to human needs. However, the degree of variation allowed in a manufacturing system can be defined for variables. In this case, the system can be defined as the manufacturing process. In most cases, the variables will be compared to telemarketing produced on a procedure. By measuring the variables of an operating procedure and allowing for variation, how far the operations will vary from the target can be determined.

The columns in Table 3.10 are labeled similarly to those for attributes. If the procedure produces a part that varies ±0.05 minutes, it would be considered minor. When a telemarketing call varies ±0.25 minutes, however, it would be considered a critical risk and rejected.

Table 3.10 is similar to a service development specification. Most engineers define services in terms of numbers. The major difference is that quality specifications must reflect real-world needs. Some service developers do not take into account the amount of variation that occurs in service systems. Quality specifications reflect the process variation and customer needs.

It is not unusual to see a conflict between service development specifications and quality specifications. It occurs because many developers do not follow their designs through the service process implementation. Because of this, they may not understand the relationship between actual customer needs and serviceability.

Operating Conditions

Operating conditions describe the environment in which the service is to run. In effect, they define the state of nature the service will be exposed to in its normal use. This includes office configuration, temperature, and equipment. A list of operating conditions a telemarketing service might experience is shown in Table 3.11.

Table 3.11 Service Operating Conditions for Telemarketing

Item	Condition	Nominal unit	CR	MA	MI
1	Employee experience	4 years	1	2	3
2	Room temperature employee	72 degrees	≤64	≤66	≤70
3	Room temperature equipment	65 degrees	<50	<52	<56

It is important to list as many conditions the service might possibly be exposed to—from employee experience to room temperature. As shown in Table 3.11, operating conditions are described in the same manner that variables are written. Here, too, each condition must be associated with the five basic human needs.

Some of these conditions are defined by the service development group. Usually, however, the developers do not have the knowledge to define the environment except in the simplest terms. It is up to reliability engineers to define these conditions. Reliability engineers are part of the quality profession.

Application

External needs are defined by describing attributes, variables, and operating conditions. Each of these categories has a risk associated with it, ranked critical, major, or minor. These descriptions can be used to compare how well the output of a system does what was required by the external needs. The performance ratio is the system efficiency.

The characteristics can be used for developing the first or second house of a quality function deployment chart. Any service may provide several key "voice of the customer" or "counterpart" characteristics. Such analysis would be for long-term, statistically significant characteristics.

One way to determine the significance of a characteristic is by developing a design-of-experiments model. If such a model is developed according to Taguchi methodology, the environmental factors can be isolated. The attributes are sorted into conforming and nonconforming. Finally, the variables are handled with the appropriate Taguchi model.

The ISO 9000 standards do not require the use of any of the service analysis tools. However, the standards do require at least some, or all, of these tools if they are required by the particular business. If the service analysis tools are used, the standards require documentation of the way they are used in ISO 9000 Section 4.20 and the results in Section 4.17.

DEFINING THE PROCESS

Another vital step toward developing a stable quality system and achieving ISO 9000 registration is defining the system used to produce the service. The starting point should be an examination of the service methods used in a particular organization. It is necessary to evaluate the operation in a top-down manner in order to understand the decisions, methods, and procedures used to make a service. The following are several tools used to describe a basic manufacturing system.

Organizational Chart

The first step in defining a system is to describe the decision-making authority and the responsibility of each person in the system. This is done by creating an organizational chart. The chart depicts the lines of authority and responsibility in the system. A job description of each position should accompany the organizational chart. Such a chart should be made for the entire company. In order to meet the provisions of ISO 9000, Section 4.1.2 Organization, a chart of the positions that interrelate with quality is often included in the quality manual (this is required under Section 4.2.1 Quality System, General).

A typical organizational chart is shown in Figure 3.2. The element of each person and major elements of the departments are determined from this chart. Interdepartmental comparisons can also be made.

A basic outline for job descriptions that must be attached to the organizational chart is provided in Table 3.12. Under ISO 9000 Section 4.1.2.2, these descriptions are one way of ensuring that proper scope and depth are given to the appropriate people.

Organizational charts will help to determine what emphasis the company places on various operations. One may discover that the quality department is under the manufacturing department. Such a relationship has been previously described as inefficient because it is equivalent to placing the accounting or personnel departments under manufacturing. Another interesting facet of organizational charts is the management level of each of the departments. It might be discovered that all people reporting to the president of the company are vice presidents, except for the quality manager. Such an arrangement clearly

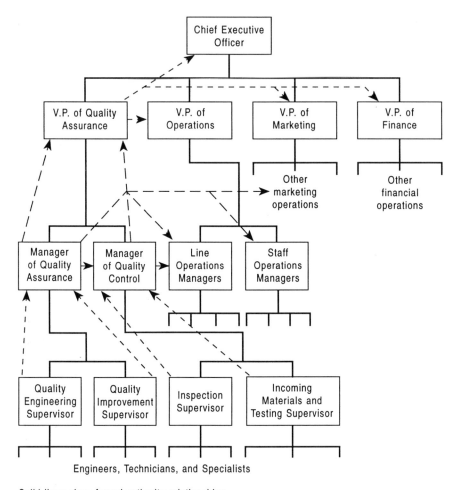

Solid lines show formal authority relationships.
Dotted lines show semiformal advisory relationships.

Figure 3.2 An "Ideal" Quality Organization Chart (*Source:* Evans, J. and W. Lindsay (1993). *The Management and Control of Quality.* West: Minneapolis, Minn., p. 189. Reproduced with permission.)

demonstrates the lack of emphasis decision-makers in the company have given to quality.

Completed organizational charts should help determine where problems originate. They will also help evaluate the system that has been established to determine its efficiency.

Table 3.12 Outline for Job Descriptions

Term	Description
Education	The highest level of education required to perform the work
Experience	The highest amount of experience required to perform the work
Job description	Tasks, duties, activities, performance standards, and decisions
Job specification	Abilities, skills, and knowledge needed to perform the job

Procedural Flow Diagrams

Another chart used to define a system is a flow diagram. This chart shows the flow of materials or information (documents) throughout the operations in the system, from startup to finished service. One good use of the chart is to trace materials or service flow (useful in analyzing most ISO sections, including 4.6. Purchasing, 4.8 Product Identification, 4.10 Inspection and Testing, 4.13 Control of Non-Conforming Product, 4.14 Corrective and Preventive Action, and 4.15 Handling, Storage, Packing, Preservation, and Delivery).

Figure 3.3 shows a flowchart. In Figure 3.4, the process flows from top to bottom. This flowchart uses a grid structure to trace the flow of work activities and information. The process flowcharts are commonly referred to as *system flowcharts* or *flow process charts*. Meyers (1992), an expert on the technique, says:

> The flow process chart is the most complete of all the techniques, and when completed the technologist will know more about the plant's operation than anyone in that plant.

Basic symbols provide a pictorial representation of activities in the process flow of a business. Although in many cases the chart can be created by drawing a circle to represent each operation and then

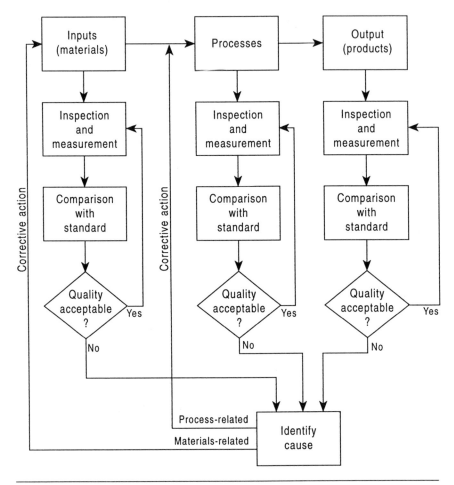

Figure 3.3 Generic Process Quality Control System (*Source:* Evans, J. and W. Lindsay (1993). *The Management and Control of Quality.* West: Minneapolis, Minn., p. 212. Reproduced with permission.)

connecting circles in sequential order, some basic process flow symbols from various sources are shown in Table 3.13.

Like the organizational chart, a brief description of each of the processes defined in the flowchart should be included on a separate page. It is important not to clutter the chart with descriptions and explanations. If too much information is placed on the chart, it will

Fred Meyers & Associates												PROCESS CHART	
DATE: 5/6/85												PAGE 1 OF 1	

☐ PRESENT METHOD   ☒ PROPOSED METHOD

PART DESCRIPTION: CASTINGS

OPERATION DESCRIPTION: PREPARE CASTING FOR PACKOUT

SUMMARY	PRESENT		PROPOSED		DIFF.		ANALYSIS
	NO.	TIME	NO.	TIME	NO.	TIME	WHY WHEN
O OPERATIONS	8	2452	4	1315	4	1137	WHAT WHO
⇨ TRANSPORT	9	779	2	0	7	779	WHERE HOW
☐ INSPECTIONS	—						
D DELAYS	3				3		FLOW
▽ STORAGES	2				2		DIAGRAM
DIST. TRAVELED	1420 FT.		240 FT.		1180 FT.		ATTACHED

STUDIED BY: FRED MEYERS   (IMPORTANT)

STEP	DETAILS OF PROCESS	METHOD	OPERATION	TRANSPORT	INSPECTION	DELAY	STORAGE	DISTANCE IN FEET	QUANTITY	TIME IN HRS/UNIT TMU	COST PER UNIT	TIME/COST CALC.
1	UNLOAD	HAND	●	⇨	☐	D	▽			31	.00217	@$7.00/HR
2	MOVE TO PUNCH	CON-VEYOR	O	➡	☐	D	▽	40'		34.3		NO COST
3	PUNCH WINDOW HOLE		●	⇨	☐	D	▽			642	.04494	
4	HANG		●	⇨	☐	D	▽			321	.00225	
5	MOVE TO PAINT	OVER HEAD	O	➡	☐	D	▽	200'		FREE		
6	PACKOUT		●	⇨	☐	D	▽			321	.00225	
7			O	⇨	☐	D	▽					
8			O	⇨	☐	D	▽					
9			O	⇨	☐	D	▽					
10			O	⇨	☐	D	▽				.0516	
11			O	⇨	☐	D	▽					
12			O	⇨	☐	D	▽					
13			O	⇨	☐	D	▽					
14			O	⇨	☐	D	▽					
15			O	⇨	☐	D	▽					
16			O	⇨	☐	D	▽					
17			O	⇨	☐	D	▽					

Figure 3.4 Flow Process Chart (*Source:* Adapted from Meyers, F.E. (1992). *Motion and Time Study.* Prentice Hall: Englewood Cliffs, N.J., p. 51.)

Table 3.13 Procedural Chart Symbols

⬁	**Document symbol** (computer science)	Used to describe any input or output that is a paper document
◯	**Operation symbol** (industrial engineering)	Shows an activity or conditions of a service element
◇	**Decision symbol** (computer science)	Depicts operations that determine which of two or more paths will be followed
●	**Connector symbol** (electrical engineering)	Used to document continuation of the flow to another point
◡	**Terminal symbol** (electrical engineering)	Shows the start or end of a process or subprocess
▽	**Storage symbol** (plant layout)	Depicts the activity of storing information or materials
☐	**Inspection symbol** (industrial engineering)	Used to represent a test or physical inspection
◗	**Delay** (industrial engineering)	Depicts a delay for information

become unreadable and defeat its purpose, which is to illustrate how operations flow within the system. Any notes can be included on a description attached to the chart. If readers need additional detailed information, they can find it on the attached pages. An easy rule to remember is KISS: keep it short and simple.

Procedural Analysis

The symbol-enhanced flowchart can also be used for a detailed procedural analysis, as well as for documenting the process details of operations (useful in meeting the requirements of Section 4.9 Process Control and assisting in documenting 4.18 Training). To define a system is to describe the steps used to perform the operation. Process chart drawing requires observation of the procedures being completed.

When certain that all the steps have been seen, a detailed step-by-step analysis of each action taken must be written. Finally, the appropriate symbols can be placed on the chart.

Performing a basic analysis of these procedures does not require technical knowledge. The purpose of the diagram is to enable nontechnical people to understand in the simplest terms how the process is completed. To accomplish this, the diagram only shows those parts that actually affect the service and what it produces. Again, it is not necessary to have design knowledge to understand procedure operations. In fact, technical people may overlook the purpose of the procedure and view it as a finite technical system devoid of its intended use.

In addition to providing information for the ISO 9000 system, another reason for conducting procedural analysis is that service managers may not fully understand how their procedures operate. In fact, some system developers may not know how their designs will operate in actual practice. The best resource is the employee who has day-to-day responsibility for a particular operation. The amount of information an experienced employee can provide is substantial.

Figure 3.5 shows a procedure analysis for a telemarketing routine. Only those operations which are critical to the service are analyzed.

Procedural Flow Diagrams and ISO 9000

When describing a system for ISO 9000 registration, three analytical techniques must be utilized. They will define the quality organization, trace the procedures and materials, and detail the activities each employee must perform. Each of these charting techniques includes additional information in an attached detailed description. The techniques are simple enough for the average employee, such as a quality circle member or a performance team member, to understand, document, and explain.

Other techniques, such as activity analysis (which includes right- and left-hand charts, line balance charts, and predetermined time systems), are often used, but require specialized training.

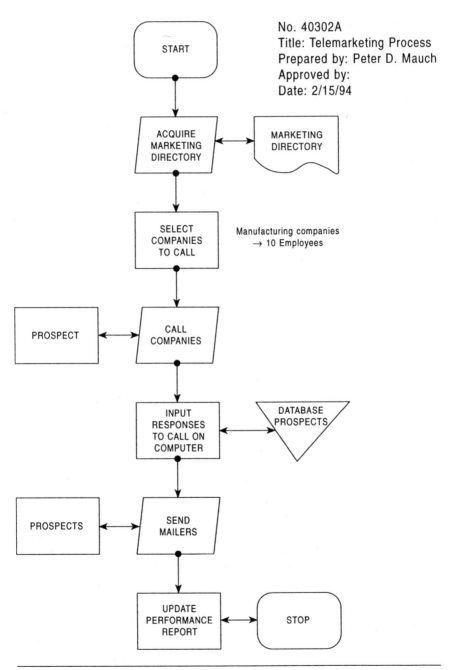

No. 40302A
Title: Telemarketing Process
Prepared by: Peter D. Mauch
Approved by:
Date: 2/15/94

START

ACQUIRE MARKETING DIRECTORY

MARKETING DIRECTORY

SELECT COMPANIES TO CALL

Manufacturing companies
→ 10 Employees

PROSPECT

CALL COMPANIES

INPUT RESPONSES TO CALL ON COMPUTER

DATABASE PROSPECTS

PROSPECTS

SEND MAILERS

UPDATE PERFORMANCE REPORT

STOP

Figure 3.5 Procedure Analysis for Telemarketing

PROBLEM SOLVING

It is a fact of life that problems exist. Much of the philosophy of the ISO 9000 standards and many of the sections deal with how a company analyzes and documents its problem-solving techniques. The appropriate ISO 9000 sections include 4.2.3 Quality Planning, 4.4 Design Control, 4.10 Inspection and Testing, 4.12 Inspection and Test Status, 4.13 Control of Non-Conforming Product, and 4.14 Corrective and Preventive Action.

Although some people, to their credit, actively pursue problem solving, solution of the observed problem is not necessarily enough. Sometimes solving a system problem only masks the root of the problem. An analogy would be people fighting a fire in the same location, over and over again, while overlooking the reason why the fire continues to recur.

The whole purpose of problem solving is to get to the root cause. If the *cause* of the problem is discovered and solved, there is no need to find ways of dealing with it again. This may mean changing the way a task is performed to make it work better. If the method is not changed, the problem still exists. The reason many companies go out of business is that they do not change quickly enough to satisfy their customers' needs.

Included in any ISO 9000 system should be the means to evaluate a problem and give direction toward change (Section 4.14 Corrective and Preventive Action). The change must be committed to by the decision makers (Section 4.1.2.1 Responsibility and Authority). Without their commitment, solutions will not be implemented. This usually happens in companies that are complacent. The result is a poor quality service.

Problem-Solving Techniques

The technique for solving problems involves seven steps:

1. Identify the problem

2. Gather data

3. Analyze the data

4. Develop solutions

5. Develop a plan

6. Implement the plan

7. Make corrections to the plan

It is necessary to perform each step in order to successfully solve a problem. Each of these steps is discussed in detail here in order to understand of its importance.

Problem Scenario

You are the quality manager in Company ABC. It has been brought to your attention by the company owner that one of the procedures is running 28% deficient. The procedure produces telemarketing services and the last service report showed that 55 out of 200 telemarketing calls were deficient. A list of the defects found by a supervisor is shown in Table 3.14.

Table 3.14 Defect List

Defect description	# found
Lost calls	15
Complaints	4
No follow up	9
Disconnects	22
Missing scripts	3
Damaged order forms	1
No response	1
Total	55

This deficiency rate is a major concern because it represents the company's ability to provide service to a new customer, which will affect a substantial part of the telemarketing business.

It may be tempting to concentrate only on the defects that have the highest frequency as indicated on the list. Actually, the first test is to identify the problem.

Identification of the Problem

In order to identify the problem, the list in Table 3.14 must be organized in some way. This is accomplished by ranking the defects from the most frequently occurring to the least, as shown in Table 3.15.

Table 3.15   Sorted Defect List

Defect description	# found	%	Σ%
Disconnects	22	40	40
Lost calls	15	27	67
No follow up	9	16	83
Complaints	4	7	90
Missing scripts	3	5	95
Damaged order forms	1	2	97
No response	1	2	99
Total	55	99	—

In Table 3.15 there are two additional columns labeled percent and sum of percent. The values in the percent column were determined by dividing the quantity of the defect by the total number of deficiencies. For example, to find the percent of disconnects, divide 22 by 55, which yields 0.40 as a result. Multiply 0.40 by 100 to get 40%. This is done for each deficiency in Table 3.15 and placed in the column labeled percent.

In the column labeled sum of percent, the values in the five columns were added together. For example, the value for lost calls was found by adding 40 and 27, which sums to 67.

Ranking the defects allows a better understanding of the problem. This a vital step in identifying the problem. The nature of any problem is that it will exhibit a number of symptoms. The defects listed in Table

3.15 are only the symptoms of the real problem. The key is to isolate the symptoms that led to the cause of the problem. By ranking the defects, the vital few symptoms have been isolated from the trivial many.

The next step is to determine which of these defects represent the vital few that need to be investigated. In the following example, the defects are evaluated as a group in order to get a clearer picture.

EXAMPLE: Calculating Percentage Defect

What defect represent 40% of all the deficiencies?

Step 1   Disconnects   = 22
              Total defective = 55

Step 2   Percent = Disconnects/total defects
                     = 22/55
                     = 0.40
                     = 0.40 × 100
                     = 40%

Step 3   40% of the deficiencies are disconnects

EXAMPLE: Grouping Defect Sources

Which deficiencies represent 83% of all the problems?

Step 1   From percent column:
              Disconnects   = 0.40
              Lost calls   = 0.27
              No follow up = 0.16

Step 2   Σ% = Disconnects + lost calls + no follow up
                 = 0.40 + 0.27 + 0.16
                 = 0.83
                 = 83%

Step 3   Disconnects, lost calls, and no follow up represent
              83% of all the deficiencies.

In the second example, the solution could have been arrived at by looking at the column labeled sum of percent in Table 3.15 and finding the value 0.83. Normally, one would look at the ranked list to find a value between 80 and 95% in order to isolate the critical symptoms. If the cause of the disconnects, lost calls, and no follow up is found, 83% of the deficiencies will be eliminated.

From these data, the problem symptoms can be identified as disconnects, lost calls, and no follow up. This defines the problem in a way that points the direction toward a useful solution. Without this approach, the root cause of the problem may never be discovered. This would result in the symptoms recurring again and again.

Gathering Data

The next step in solving problems is to gather information specific to the vital few symptoms. It is important to concentrate solely on gathering data specifically on the vital few symptoms, not on the trivial many.

With respect to the symptoms listed in Table 3.15, the following additional information was found:

1. Disconnects occur in the morning
2. Lost calls are caused by disconnects
3. No follow up is the result of lost calls
4. The average call time is 4.5 minutes
5. The average time of a call varies ±5 minutes

The information listed is specific to the symptoms under investigation. It cannot be overemphasized that the data collected must be related to the symptoms being researched. Otherwise, the data are worthless.

Analysis of the Data

Analyzing data requires statistical and technical knowledge. To relate the data to a cause requires scientific understanding of the process. Technical knowledge alone is not enough because an understanding of natural process variation is essential. This requires a good understanding of the statistical method in describing states of nature.

From the information gathered for the symptoms in Table 3.15, a common characteristic can be derived. From the information given in the preceding paragraphs, it appears that disconnects and call time are characteristics common to the symptoms.

Developing Solutions

In the preceding step, disconnects and lost calls were found to be common to the problem symptoms. One solution that could be developed for this example is that disconnect is the problem causing all the deficiencies (symptoms) to occur.

Most problems are not this obvious. It is possible to have many different solutions for a problem. These solutions can be developed by getting a group of people together and brainstorming about possible solutions. It will be necessary to rank order these solutions based on the amount of resources needed to implement their correction. This will help in the next step of developing a plan.

Developing a Plan

Based on the solutions, a plan must be established to allocate the required personnel and resources necessary. An action plan requires a list of the steps needed to be performed and should include the length of time to complete each step. An activity list for a plan to solve the problem in Table 3.14 is provided in Table 3.16.

In Table 3.16, each activity and the amount of time required is listed. Additionally, the column labeled prerequisites lists those activities that need to be performed before proceeding to the next step in the plan. The plan can be further developed using a timeline chart, as shown in Table 3.17.

Table 3.17 shows a visual representation of the time and steps required to complete the plan. The plan could be broken down further by the amount or cost of the plan. When a plan is costed out, three

Table 3.16 Activity List

Activity	Duration (in weeks)	Prerequisites
1. Arrange for test equipment	2.0	None
2. Arrange for staff	1.0	1
3. Allocate service downtime	2.0	2
4. Perform tests	0.5	3
5. Correction and summary	3.0	4

elements must be estimated: direct labor cost, direct material cost, and allocated overhead cost.

Direct labor costs are the hourly costs of the people who must work on the problem. The hours can be determined from the activity list, and an hourly rate can be applied to arrive at the dollar amount.

Direct material costs are the costs of equipment rentals and spoilage created by the testing. It will be easy to acquire the rental rates for equipment, but material spoilage must be approximated.

Allocated overhead cost is the fixed cost associated with doing business. This is the expense the company would have to pay even if it did not make any money. Examples of allocated overhead costs are rent, utilities, and licensing fees. For general purposes, 65% of the combined direct labor and direct material cost can be used for allocated overhead costs, as a rule of thumb. Assuming that the direct labor cost

Table 3.17 Timeline Chart

Activity	1	2	3	4	5	6	7	8
1. Test equipment	XXXXXXX							
2. Staff		XXXXXXX						
3. Downtime			XX					
4. Testing			XXXXXXXX					
5. Corrections					XXXXXXXXXXXXXX			

for this plan is $50,000 and direct material costs are $10,000, the total for direct labor and materials is $60,000 ($50,000 + $10,000). Allocated overhead cost will be 65% of $60,000, or $39,000. Therefore, the total cost for the plan is $99,000. This is the cost to implement the plan.

Implementing the Plan

To correct the problem, it is important to implement the plan that has been developed. Many solutions have never been found simply because the plan was never implemented.

Corrections to the Plan

Since no plan can be foolproof, it is necessary to modify the plan if the correct results are not obtained. Here, too, many plans have not solved problems because management refused to change the plan when the desired results were not obtained. It may be necessary to revise the entire plan and look for other solutions.

INSPECTION AND TESTING, CALIBRATION, AND QUALITY RECORDS

Inspection and Testing

Although there seems to be a trend away from inspection or, at least, product sampling inspection, the ISO 9000 standards anticipate a traditional three-stage inspection program: incoming materials, in-process inspections, and final inspection. Section 4.10.1 Inspection and Testing requires the company to maintain documented procedures for inspection and testing activities to verify that the specified requirements for the product are met. In the case of receiving (Section 4.10.2), such requirements were established in the purchase agreement (Section 4.6 Purchasing), and data from such inspections will be used to evaluate subcontractor performance. In the case of final inspection (4.10.4), all activities must be completed and documented before material is processed out of production. The requirements for inspection are developed according to the quality plan (Section 4.2.3) and relate to 4.3 Quality Review or 4.4 Design Control. Care of the verified materials after inspection is required by 4.15 Handling, Storage, Packaging, Preservation, and Delivery and later by 4.19 Servicing.

Sampling plans must be completely defined in the inspection documents. That is, if MIL-Q-105E is called for, operators must be thoroughly familiar with all phases of the specification. Simplified tables of commonly used values are often developed and issued as a controlled quality document (Section 4.5).

Materials found to be nonconforming (Section 4.13 Control of Non-Conforming Product) must be isolated and stored according to procedures that will prevent unintended use. Disposal of the product must be in accordance with procedures. Records (Section 4.16) of such nonconforming materials will be kept and the customer notified if a product does not meet specified requirements. Corrective and preventive action (Section 4.14) procedures will be followed to eliminate the source of the nonconformity. If not resolved, the problem will become subject to 4.1.3 Management Review.

Calibration

Calibration of test equipment to primary standards is required under Section 4.11.2.d and 4.11.2.e Control Procedure. Guidance may be found in ISO 10012. However, there are three calibration areas that must be evaluated in particular. First, many prospective registrants do not include gauges and meters on process equipment. If it is used in controlling the process, it must be calibrated. It is particularly critical to calibrate gauges when used for controlling special processes (Section 4.9). Second, software must have backup security and tests. The calibration must occur at documented routine intervals, and records must be maintained. Finally, the ability of the users to accurately and reliably use the test equipment must be assessed. Documentation must be maintained.

Quality Records

Records, according to 4.16 Control of Quality Records, must be maintained to demonstrate conformance to the specified requirements. Procedures must exist to identify, collect, index, access, file, store, maintain, and dispose of records. Electronic or other media may be used, provided the security issue is controlled.

MANAGEMENT RESPONSIBILITY, AUDITING, CORRECTIVE AND PREVENTIVE ACTION

Management will review the quality system at defined intervals (Section 4.1.3). Records of these meetings must be maintained (see Section 4.16). At these meetings, management is expected to assess performance against ISO 9000 and stated quality policy (see Section 4.1.1). The results of the 4.13.2 Non-Conforming Product Review and Disposition must be reviewed to ensure that appropriate policy is being followed. Any ensuring that the policies controlling 4.14.2 Corrective Action and 4.14.3 Preventive Action are both sufficient and implemented must be evaluated and the results of audits reviewed. Responsibility for all action items should be completed at the meeting.

The Internal Quality Audit (Section 4.17) is the primary means of verifying the effectiveness of the system. It supplies the information for management review (Section 4.1.2.4). Results of the audit and timely corrective action are to be effected and published. A plan must be established to schedule and implement the audit. Personnel without direct responsibility for the audited area are to conduct the audit. ISO 10011 contains guidelines for conducting an audit.

The greatest misunderstanding about ISO 9000 relates to Section 4.14 Corrective and Preventive Action. This section is not merely an extension of product inspection procedures. It refers to all nonconformities found in the process and requires that such nonconformities be corrected or specifications revised. It also involves analyzing all quality results to prevent (improve) nonconformities. This is the basis for improvement. The methodology for collecting proof that appropriate preventive actions are being taken requires statistical techniques. Such techniques must be documented in Section 4.20. This is perhaps the most significant change in the 1994 standard revision.

ENDNOTE

Meyers, F.E. (1992). *Motion and Time Study*. Prentice Hall: Englewood Cliffs, N.J., p. 44.

CHAPTER 4

QUALITY BEYOND ISO 9000

COMMON ISSUES

In the Star Trek series, Captain Picard gives his team instructions by the message "make it so." This is typically how most companies treat quality. They hire a quality manager and tell him or her to "make it so." What is explicit in Captain Picard's message is that his team knows what his goals are. Unfortunately, few organizations have defined their objectives for quality clearly enough to provide a consistent focus for quality. In fact, when the rewards and punishments come, reacting or not reacting to quality generally has little effect except when it comes to emergencies. Think about how many people have risen in your organization by focusing on quality.

A chief executive officer (CEO) yelling to his or her officers about a quality problem has little impact on quality because it is not considered to be a major management issue. The CEO will spend very little time defining objectives for quality, reviewing quality performance, allocating sufficient resources to quality, and establishing the organizational structure to make quality happen.

Many organizations establish a quality improvement process to focus the employees on the subject of quality. In most cases, the quality improvement process results are insignificant to the investment. Failures occur because there is no quality infrastructure or the quality improvement process has not been integrated into the normal business flow.

PROBLEMS IN IMPLEMENTING QUALITY THEORY

When discussing the subject of quality, it is easy to discover wide variation in the definitions of quality terminology. This wide variation makes understanding the subject extremely difficult. Many people talk about eliminating quality departments. If they understood quality management in sufficient detail, organizations would be expanding the role of the quality department. The function of quality, which may be reduced when other functions become operational, is inspection. Unfortunately, it is not always easy to find such obsolete inspection activities. Some organizations have changed the names of departments doing such obsolete functions to more modern, popular names without changing function. Others have included the inspection function with other quality functions.

Recently the quality improvement concept has been incorrectly called *total quality management.* It is more correctly called employee involvement in quality improvement. Quality must be integrated into the way the business is run. For example, an organization implements an employee involvement program to improve quality. However, a business that does not establish quality measures by which to manage and continues to use the same measures as before will soon fail in its effort to improve quality. There might be some initial enthusiasm, but as day-to-day business pressures take over and management prioritizes its work, quality focus will end up low on the list of priorities. A manager may get fired for not acting on a financial issue or for failure to deliver a service on time. The same person may be reprimanded for not acting on quality. Which would you do first?

In these cases, organizations may think they have adopted a total quality management approach by establishing a quality improvement process. In reality, they have adopted an aspect of total quality management and have not established the infrastructure to make quality happen.

A major part of the problem with managing for quality involves the inconsistent use of quality terminology. How can quality be effectively managed if the theory and its application are difficult and confusing?

TERMINOLOGY

Quality

Quality is defined as **meeting customer needs. This** definition allows quality to be measured in terms of the customer, who may be internal or external to an organization. Quality may be measured in terms of customer satisfaction, product performance, reliability, delivery, cost, or some other suitable indicator. If one is to manage quality, one must be able to measure it.

If you replace the word *meeting* with *exceeding*, the definition has a better ring to it, but it becomes more difficult to measure and manage quality. If you purchase an automobile, what does exceeding customer needs mean? Consider the following:

1. More features for the cost

2. Longer life

3. Leather interior

4. Other features

The first step that an organization must take is to understand and define customer needs. An organization can have a competitive advantage by targeting specific needs and going beyond (exceeding) those targeted needs. However, those targeted needs are still treated internally as meeting the requirements.

Customer needs change over time, and an organization must also consider this aspect in its definition of quality. As customers continue to receive product that initially exceeded their expectations, this new level becomes the standard and eventually the customer expectation.

Market grade quality must not be confused with defects in product quality. The simplest example might be purchasing an automobile. Minivans, luxury cars, and high-mileage cars are intended to satisfy certain aspects of the marketplace. The customer would expect each car to be defect free when purchased.

Quality System

The quality system is defined as **the organizational structure, resources, policies, processes, and procedures that effectively and efficiently deliver quality.**

If a quality system is sound, it should deliver quality to customers and be effective and efficient in doing so. If someone indicates that compliance to a quality system does not provide a quality product, the question should be raised as to whether the quality system is worth implementing.

If the quality system is not efficient in delivering quality, the organization will not survive, especially if other organizations have efficient quality systems. The inefficiencies in the quality system are impacted by cost and delivery delays.

Quality Management

Quality management is defined as **management of the quality system**.

One aspect of this definition provides for the role of the quality organization and the top quality officer. That person's role is not as the chief inspector, but rather an integral part of top management with responsibility for managing the quality system. That person will be responsible for reporting on quality, auditing for quality, ensuring maintenance of the current system, and acting as a catalyst for change.

A typical background for a person to assist in implementing a quality system, such as the ISO 9000 standards, is as follows:

1. Technical degree

2. Business degree

3. Certified quality engineer or auditor, certified reliability engineer is a plus

4. Ten years in quality system engineering, with at least five years in management

Total Quality System

A total quality system is defined as **deployment of the quality system to all aspect of the organization**.

Total Quality Management

Total quality management is defined as **managing the total quality system**. It has a variety of dimensions:

1. Organizational coverage

2. Extent of the quality system elements implemented in each part of the organization

3. Degree of deployment of requirements of each quality element

Policy

Policy is defined as a **governing principle for people in an organization to follow**. Typically, a policy will define what is to be accomplished and who has authority for it. An organization may have multiple policies and multiple layers of policies.

Objective

An objective is defined as **a goal to be accomplished by a specific time**. It may cover results to be achieved or activities to be accomplished. In either case, it must be measurable and achievable. Results should be measured against the objective and reviewed by management. Management should show how their actions relate to the objective.

COMMENTS ON QUALITY STANDARDS

When American National Standards Institute Committee Z-1 reviewed ISO 9004 (ISO version of 9004) to determine if the standard was complete, the following issues were pointed out:

1. Quality information system was not included.

2. Quality improvement was not covered.

3. The standard was manufacturing oriented.

4. The structure of the standard provided misleading emphasis. Production was covered in multiple elements, while the design clause was covered in one element.

The other aspect of this standard stressed the importance of standardized vocabulary. Leading experts had difficulty agreeing on terminology. Think about how confusing this must be to people who are not familiar with the subject. However, these experts could readily agree on what constitutes a good quality system.

As a result, the ANSI Z-1 developed a U.S. draft of a Total Quality Management Standard to represent a consensus of leading experts. An updated version of that input is provided here.

STANDARD FOR TOTAL QUALITY MANAGEMENT

INDEX

0 INTRODUCTION

A primary concern of any organization should be the quality and consistency of its services, processes, and products.

In order for services to be successful, an organization must offer products or services that consistently:

a. Meet a well-defined need, use, or purpose

b. Satisfy customers' expectations

c. Comply with applicable standards, specifications, and with statutory (and other) requirements of society

d. Are safe to use

e. Are made available at competitive prices

f. Are constantly improved

The quality system should provide proper confidence that:

a. The system is well understood and effective

b. Products and services are designed to achieve and maintain the required level of quality and to be robust to variations in the environment in which they are used

c. The products or services actually do satisfy business requirements and customer expectations and the requirements of society

d. Emphasis is placed on problem prevention rather than dependence on detection after occurrence

e. The quality improvement process is effectively working

The quality management system should be deployed throughout the business and be oriented toward designing quality into the service, process, or product and then reducing, eliminating, and preventing quality deficiencies in the production and delivery of the product or service.

The approach used in this standard is to describe a product or service life cycle model (customer needs identification; service, product, and process development; and operation) along with a support structure (utilization of people, management leadership, and quality information and results) and continually improve the efficiency and effectiveness of business processes. This is illustrated in Figure 4.1.

Management is ultimately responsible for establishing the quality policy and for decisions concerning the initiation, development, imple-

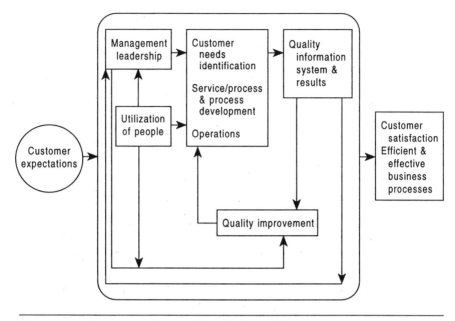

Figure 4.1 Product or Service Life Cycle Model

mentation, maintenance, and evaluation of the quality management system.

Senior executives should provide the leadership, involvement, and support to the quality management system. They should organize the technical, administrative, and human factors affecting quality to achieve constancy of purpose.

The quality system should be structured to continuously draw upon and be driven by feedback from the ultimate user of the product or service and from those who work within the system.

Management should continuously monitor the quality system in order to make the necessary adjustments, changes, and advancements needed to ensure that the system's products and services continue to meet the stated requirements, satisfy its customers, and continually improve.

The system requires objective evidence in the form of information and data concerning the quality of the system and the quality of the company's products and services.

1 SCOPE, PURPOSE, AND APPLICATION

1.1 Scope

This standard describes a set of practices for quality management and quality systems.

To describe these quality management practices, the standard utilizes a group of seven categories. These seven categories provide the basis for the organization of the standard and are titled:

1. Management leadership
2. Quality information system and results
3. Utilization of people
4. Customer needs identification
5. Service, product and process development
6. Operations
7. Quality improvement

1.2 Purpose

This standard provides guidelines for developing, implementing, maintaining, and improving a quality management system.

1.3 Application

The quality system should be structured and adapted to the organization's particular type of business and should take into account the appropriateness of categories as outlined herein.

Each element within a category should be reviewed to determine its applicability to a given business.

In order to consistently achieve maximum effectiveness and to satisfy customer expectations, it is essential that the quality management system be (1) deployed throughout the business and (2) integrated with business strategy.

Total quality management and/or quality systems is the application of the model specified in this standard to all aspects/parts of the organization.

2 NORMATIVE REFERENCES

IEC 300: Reliability Management

ANSI/IEEE 730: Software Quality Assurance Plans

ANSI/ASQC Q1: Generic Guidelines for Auditing of Quality Systems

ANSI/ASQC Ml-1987: Calibration Systems

ANSI/ASQC (TBD): Guide for Formal Design Review

ANSI/ASQC Q90

ANSI/ASQC Q91

ANSI/ASQC Q92

ANSI/ASQC Q93

ANSI/ASQC Q94

3 DEFINITIONS

Definitions are provided in the following standards:

- ANSI/ASQC A1: Definitions, Symbols, Formulas, and Tables for Control Charts
- ANSI/ASQC A2: Terms, Symbols, and Definitions for Acceptance Sampling
- ISO 8402-1986: Quality—Vocabulary

For purposes of this standard the definitions in ISO 8402 may be interpreted in this standard as follows:

- **Quality Management:** Activities of overall management that establish, manage, and improve the quality of the quality system.

- **Quality System:** The organizational responsibilities, resources, processes, and procedures needed to effectively and efficiently deliver quality.

The use of the word **total** in front of quality management or system refers to their deployment to all aspects of the organization.

4 MANAGEMENT LEADERSHIP

4.1 Quality Policy

The management of an organization should develop and state its quality policy. This policy should be documented, comprehensive, and support the overall business/organization strategy with expected behavior levels of performance to meet the requirements of this standard and the requirements of society.

The basis of the quality policy should be prevention rather than detection and correction of a problem.

Management should take all necessary measures to ensure that its quality policy is understood, implemented, and maintained.

4.2 Strategic and Short-Term Quality Objectives

Management should define strategic and short-term objectives pertaining to key elements of quality, such as fitness for use, performance, reliability, safety, improvement, and customer satisfaction.

The calculation and evaluation of costs associated with poor quality should always be an important consideration, with the objective of minimizing quality losses to achieve efficient and effective processes.

Appropriate levels of the organization, where necessary, should define specialized quality objectives consistent with quality policy and objectives.

4.3 Structure of the Quality System

4.3.1 Responsibility and Authority

Activities contributing to quality, whether directly or indirectly, should be identified and documented, and the following actions taken:

a. General and specific responsibilities should be explicitly defined.

b. The lines of authority and communication should be defined and open communication channels should be emphasized.

c. Interface control and coordination measures between different activities should be defined.

d. Responsibility and authority delegated to each activity contributing to quality should be clearly established; authority and responsibility should be sufficient to attain the assigned quality objectives with the desired efficiency.

e. Management may establish a support function to assist them in managing quality; the support structure should be independent of the activities reported on.

4.3.2 Resources and Personnel

Management should provide sufficient and appropriate resources essential to the implementation of quality policies, the achievement of quality objectives, and to develop and maintain a culture that supports these objectives. Resources include financial, people, equipment, and material.

4.3.3 Organization Emphasis

The quality system should be organized in such a way that all activities affecting quality are integrated into a common focus of efficiently and effectively meeting customer needs.

The quality system should emphasize robust design and preventive actions, while not sacrificing the ability to respond to and correct failures should they occur.

4.3.4 Culture

Management should develop a culture that promotes the following:

a. A positive environment that is stable and secure

b. Respect for the individual while encouraging teamwork and managing diversity

c. A global perspective through knowledge of philosophies, policies, and practices external to the company

d. Open communication

e. Reporting of problems without fear of retribution

4.4 Senior Management Review and Action Plan

Management reviews should consist of well-structured and comprehensive evaluations which include:

a. Audit results centered on various categories of the quality system

b. The overall effectiveness of the quality system in achieving stated quality objectives

c. Considerations for updating the quality system in relation to changes brought about by new technologies, quality concepts, market strategies, and social or environmental conditions or requirements of society

d. Review of training objectives and performance criteria

e. Customer feedback

f. Other relevant sources of information

The reviews should be carried out by appropriate members of senior management, assisted if desired by competent, independent personnel reporting directly to senior management. Formal senior management review should take place no less frequently than quarterly. Documentation of the review and action plan should be maintained.

4.5 Managing the Quality System

4.5.1 Responsibility

Executive management has overall responsibility for quality. Management may establish a support function to assist them.

The people responsible for supporting management should have knowledge and skill in quality principles and practices.

4.5.2 Quality Plans

A documented quality plan should be prepared to manage new service, product, or process development; to manage the ongoing business; and to improve quality. The quality plan should include:

a. The quality objectives to be attained

b. The specific allocation of resources, responsibilities, and authority

c. The specific business processes, procedures, and methods to be applied

d. The specific training requirements and performance criteria

e. Process control and how it is to be achieved, including suitable testing, inspection, examination, and audit programs at appropriate stages (e.g., design, development, service delivery, production)

f. A method for changes and modifications to the quality plan as projects proceed

g. Quality and reliability engineering plans

h. Qualification requirements

i. Requirements regarding impact of services, processes, or products on the environment

j. Regulatory requirements

k. Other measures necessary to meet objectives in accordance with the quality management system

l. A method for measuring customer satisfaction

The organization or organizations required to implement the plan should agree to it and obtain concurrence of other impacted organizations.

4.5.3 Documentation of the System

4.5.3.1 Quality Policies and Procedures

The quality system should be documented in a systematic and orderly manner in the form of written policies, processes, and procedures. Such documentation should ensure a common understanding of quality policies and procedures.

The quality system should include adequate provision for the proper identification, distribution, collection, and maintenance of all quality documents and records. However, care should be taken to limit documentation to the extent pertinent to the application.

The quality system should require that sufficient documentation be available to demonstrate the attainment of the required quality and the effective operation of the quality management system.

4.5.3.2 Quality Manual

The typical form of the main document used in drawing up and implementing a quality system is a quality manual.

The primary purpose of a quality manual is to provide an adequate description of the quality system while serving as a permanent reference in the implementation, maintenance, and evaluation of that system.

Provisions should be established for making changes, modifications, revisions, or additions to the contents of the quality manual. Such changes should be appropriately documented.

The documentation relating to the quality management system may take various forms, including the following:

a. Organizational quality manual

b. Divisional or office quality manuals

c. Specialized quality manuals (e.g., design, procurement, project, business process instructions, work instructions)

4.5.3.3 Document Control

A method for removing and/or disposing of out-of-date documentation related to quality should be established to ensure that the correct documentation is being used.

The following are examples of the types of documents requiring control:

- Quality manual
- Drawings
- Requirements
- Specifications
- Validation procedures
- Personnel policy and procedure manual
- Business process descriptions
- Schedules
- Work instruction
- Quality assurance procedures
- Regulatory requirements including impact on the environment
- Safety requirements

4.5.4 Auditing the Quality System

Audits should be performed to determine whether various elements within a quality management system are efficient and effective to satisfy customer needs and achieve quality objectives. For this purpose, appropriate quality requirements covering the system, product, service, and process should be documented and maintained.

Details of Quality Auditing are described in ANSI/ASQC Standard Q1: Generic Guidelines for Auditing of Quality Systems.

4.5.5 Quality Records and Document Retention

A quality record retention process should be established to identify, collect, index, file, store, maintain, retrieve, and dispose of quality documentation and information.

The system should require that sufficient records be maintained to demonstrate achievement of the required quality and verify effective operation of the quality management system.

The following are types of quality records requiring control:

- Measurement records
- Sales orders
- Qualification and validation reports
- Audit reports
- Material review reports
- Calibration data
- Quality information reports
- Design reviews
- Design qualification results
- Customer complaint reports
- Purchase orders
- Employee records
- Inventory records
- Customer survey
- Corrective action records
- Senior management review and action plan

Quality records should be retained for a specified period for analysis in order to identify quality trends that demonstrate system effectiveness and the need for improvement.

Quality records should be protected from damage, loss, and deterioration due to environmental conditions.

Appropriate subcontractor documentation should be included. All documentation should be legible, dated (including revision dates), clean, readily identifiable, and maintained in an orderly manner. Data may be hard copy or stored in a computer. Legal requirements should be considered.

5 QUALITY INFORMATION SYSTEM AND RESULTS

5.1 Design of Quality Measurements

5.1.1 Purpose

Measurements should be established, analyzed, and reported to ensure the effective management of the business.

Measurements should also provide for the continual evaluation of the operation to identify and actively pursue opportunities for improvement.

To implement such evaluations management should establish and maintain an information system for the dissemination of data from all relevant collections, analyses, and sources.

Management should assign responsibility for the management of the information system.

5.1.2 Identify Measurements to Be Made

Measurements should be established and reported for appropriate aspects of the business. Data should be available from measures of the operation by means of supplier assessment, customer assessment, customer complaints, requested feedback information, product/service and process assessment, and quality audits.

Analysis of these data will measure achievement of requirements and indicate opportunities for improving quality and the effectiveness and efficiency of the quality provided.

5.1.3 Measurement Collection Sources

Measurement collection sources should include the following:

- Quality system audits
- Quality plan reviews
- Process audits
- Process and product reviews
- Product testing and inspection processes
- Quality costs systems and studies
- Employee surveys/suggestion systems
- Customer surveys, complaints, assistance requests, refunds/product returns
- Product development processes
- Quality improvement processes and projects

5.2 Use of Information

Measurements should be established for current and future goals, and analyzed on a regular basis.

5.2.1 Demonstrate System Effectiveness

Management should translate the organization's strategic goals into a set of measurable objectives. The performance measurement results should be analyzed and tracked to determine business and/or system efficiency and effectiveness.

5.2.2 Control Mechanism

The information system should be organized to provide data to all people in the organization who can impact development, control, and improvement of quality.

5.2.3 Internal Quality System Review

The quality system review should review the measurements to verify whether quality activities comply with plans and objectives to determine the effectiveness of the quality system.

5.2.4 Management Review and Action Plan

Management should utilize the quality information system for formal periodic and independent reviews of the quality system in order to determine its continuing suitability and effectiveness in implementing and managing the quality policy and achieving the quality objectives.

Particular emphasis should be placed on the need or opportunity for improvements.

5.2.5 Continuous Improvement

The information system should supply data to a process for continuously improving the quality and the effectiveness and efficiency of the complete organization. The activities relating to improvement should address short- and long-term needs of the organization.

5.2.5.1 Team Results

Members from different parts of the organization should be directed toward improving quality and reducing costs throughout the organization.

5.2.5.2 Individual Results

Personnel at all levels should be encouraged to contribute to the planning for quality, preventing deficiencies, controlling quality, and improving quality.

5.3 Type of Information

5.3.1 Customer Satisfaction

The organization should institute ongoing assessment and measurement of customer satisfaction with all aspects of its operation including products, services, delivery, billing, etc. These assessments should seek positive as well as negative reactions.

The evaluation of customer satisfaction should focus on the extent to which the service, process, or product meets the customers' needs and the likely effect on future business. A comparison should be made of customer satisfaction data with the organization's internal measures of service, process, and product quality to correlate the two measures and to take any appropriate action for quality improvement.

5.3.1.1 External

The organization should solicit and report on the levels of external customer satisfaction achieved. Consideration should be given to the following:

- Sales information and literature
- Purchasing, delivery, and billing
- Engineering services
- Installation
- Engineering technical support
- Customer support
- Repair, replacement, and/or update

- Warranty data
- System hardware/software performance
- Documentation
- Service performance

5.3.1.2 Internal

The organization should solicit and report on the levels of internal customer satisfaction achieved. Consideration should be given to the following:

- Preplanning marketing assistance
- Training
- Sales interactions
- Technical support
- Financial
- Design

5.3.2 Cost of Quality

5.3.2.1 Deployment of Quality Costs

Quality cost information should be gathered and used throughout the business and throughout the stages of the life cycle of a service/process/product, including:

- Marketing
- Design and development
- Procurement
- Manufacturing
- Testing

- Packaging, storage, and shipment
- Installation
- After-sale support
- Legal
- Finance
- Other administrative support functions

5.3.2.2 Types of Quality Cost

Quality costs are those costs incurred by an organization in order to attain, ensure, and improve quality and reliability throughout a service, process, or product life cycle. Quality costs should be broadly divided into four categories:

- **Prevention:** Activities performed in the planning cycle to ensure quality in operations (design reviews, reliability testing, customer trials, etc.)

- **Appraisal:** Activities performed to ensure that ongoing operations are meeting planned quality (testing, inspection, examination, process audits, etc.)

- **Internal Failure:** Activities that occur because of internal failure (rework, redesign, scrap, reprocessing, corrective action, etc.)

- **External Failure:** Activities that occur as a result of a failure external to the operation (warranty, returns, repairs, product recalls, liability costs, redesign, etc.)

Developing a cost of poor quality model usually uncovers the best opportunities to reduce these costs and at the same time improve quality.

The cost of activities directed at achieving appropriate quality and resultant costs from inadequate quality (cost of poor quality or nonconformance) should be identified. The cost of poor quality should be used as a driving force for improvement and should be minimized.

An organization should report on prevention costs to show that adequate resources have been deployed in the planning phase to prevent problems from occurring. Prevention costs should be expected to increase slightly with significant reductions in the cost of poor quality.

5.3.3 Performance Measurements

5.3.3.1 In-Process

Process quality control should be an integral part of the business process whether for service delivery, product realization, or business support. This includes continuous measurement and verification of in-process activities throughout all phases of the process to avoid undesirable trends, including systematic and random errors and customer dissatisfaction.

5.3.3.2 Customer Interface

Management should take steps to establish effective interaction between customers and the organization's customer contact personnel. This is crucial to the quality perceived by the customer.

Acceptance inspection and quality auditing may be used to provide feedback of the quality at the customer interface.

5.3.3.3 In Service

The organization should institute ongoing assessment and measurement of actual field performance.

5.3.4 Benchmarking

The company should actively compare its services and products against its competitors, including customer satisfaction.

Processes should be actively compared against business leaders and best-in-class organizations.

5.4 Statistical Methods

5.4.1 Applications

Modern statistical methods should be utilized to gain a better understanding of customer needs, to perform in-process control, capability studies, forecasting, or measurement of quality to assist in making decisions. Application may be for purposes such as:

- Market analysis

- Customer surveys

- Product or service design analysis

- Product reliability specification, longevity/durability prediction

- Process quality control and process capability studies

- Determination of quality levels/inspection plans

- Product failure/performance assessment/defect analysis

- Quality cost analysis

- Quality improvement trend analysis

5.4.2 Statistical Techniques

Specific statistical methods and applications available include, but are not limited to, the following:

- Design of experiments/regression analysis

- Pareto analysis/scatter diagrams/graphical methods

- Safety evaluation/risk analysis

- Cause-and-effect diagrams

- Quality control charts/sampling techniques

- Statistical survey techniques

- Product lifetime data analysis methods

- Time series analysis

5.5 Visibility of Measurements throughout the Company

Pertinent quality information should be regularly reported to and monitored by management.

The entire organization should be kept abreast of current performance and trends against company goals.

People working within a process should have process information readily visible and easily accessible to them.

5.6 Communication throughout the Supplier/Customer Chain

The company should work closely with both suppliers and customers to share information relative to quality.

The quality plan should specify what kind of information and communication should take place with suppliers and customers. The plan should include both routine sharing of business information and nonroutine situations, such as joint participation in designs, problem-solving, and concurrent process, product, and service testing.

Use of standards-based electronic data interchange for data sharing should be considered as part of the quality communication plan.

6 UTILIZATION OF PEOPLE

6.1 Selection of People

Individual people are a very significant resource in any organization. It is important that the ability of people to contribute to the objectives of the organization be maximized through appropriate personnel selection processes, including:

- Clear specification of the tasks to be performed and the objectives to be achieved, including the effect on quality

- Selection of personnel on the basis of capability to satisfy the defined specification and/or the aptitude to successfully complete job skills training

- Identification as early as possible in cases of inappropriate personnel selection
- A methodology to modify the selection criteria as requirements change and/or performance of selected personnel indicates a need to do so
- Ability to work effectively in teams

6.2 Education and Training

6.2.1 Needs Analysis

The need for education and training of personnel at all levels of the organization should be identified and a plan for providing that education and training should be developed. Particular attention should be given to the training of newly recruited employees and employees transferred to new assignments. Training requirements and expectations should be documented as a preliminary set of specifications (training brief). Workplace safety training should be provided for all levels of personnel.

6.2.1.1 Executive and Management People

Training should be provided to executive management in understanding of the quality system, together with the tools and techniques necessary for their participation in the operation of the system. Management should also understand the criteria available to evaluate the effectiveness of the quality system and quality-related costs.

6.2.1.2 Technical and Professional People

Training should be given to technical and professional people to enhance their contribution to the success of the quality system. This should not be restricted to personnel with primary quality assignments, but should include marketing, procurement, engineering, accounting, and process and product management. Particular attention should be given to training in statistical techniques such as process capability studies, statistical sampling, data collection and analysis, problem identification, problem analysis, and corrective action.

6.2.1.3 Supervisors and Workers

All production and service delivery supervisors and workers should be trained in the methods and skills required to perform their tasks, including:

- Proper operation of instruments, tools, machinery, and mechanized systems required to perform their functions

- Reading and understanding the documentation provided

- Relationship of their duties to quality

- Objectives and concepts of satisfaction of internal and external customers

- Process control, data collection and analysis, problem identification and analysis, corrective action and improvement, team working, and communication methods

6.2.2 Course Development

The course development process should translate needs from the training brief into technical specification for course materials, delivery processes, and testing. Management should establish checkpoints throughout the development process appropriate to the nature of the training. Reviews should compare the training design with the customer needs and expectations expressed in the training brief.

6.2.3 Course Qualification and Validation

The course development process should provide evaluation of the design at significant stages, including:

- Evaluation of performance under workplace conditions

- Verification that all design features are as intended
- Validation of training material and computer software

6.2.4 On-the-Job Reinforcement

The system should provide for on-the-job reinforcement to ensure that the materials learned and formal training are actually utilized in the workplace.

Periodic requalification of training should be performed in order to ensure that the design is still valid with respect to all specified requirements. The quality system should ensure that any training and field experience indicating the need for design change is fed back for analysis.

6.3 Employee Involvement

6.3.1 Quality Awareness

People at all levels should understand:

- The organization's quality policy
- The organization's quality objectives
- The role of each individual
- The individual's role in team achievement

Employees should feel that they have impact on the quality of products and services provided to their customers.

6.3.2 Opportunities for Involvement

The system should clearly delineate the opportunities for each type and classification of employee that is to be involved in the operation quality process. Types of involvement may include quality of their normal work, participation in quality improvement teams, suggestion systems, and making decisions that affect the employee's work. Management should also provide adequate systemic support to allow employees to improve quality and customer satisfaction within the scope of their work.

6.3.3 Efforts to Encourage Employees

In order to stimulate employee motivation, development, communication, and performance, management should:

- Realize the potential of every member of the organization by supporting consistent creative work methods and opportunities for greater involvement
- Establish a quality awareness program which may include introductory training courses for new employees and periodic refresher programs for long-standing employees
- Encourage participation on quality improvement teams
- Encourage contributions that enhance quality by giving due recognition and reward for achievement
- Periodically assess the factors that motivate personnel

6.4 Performance Management

Career planning and development should be implemented for all employees.

Employees should be given clear and frequent feedback by management on job performance.

Where employee performance is determined to be deficient, the supervisor and employee should attempt to determine the causes and appropriate corrective action, which could include retraining or change in job assignment.

If formal written performance appraisals are utilized, they should be intended only to confirm ongoing feedback to employees.

Caution should be exercised in the use of numerical objectives, particularly when they are used to determine compensation. Objectives should not lead to suboptimization of work processes, do not conflict with other departments, and are consistent with the overall objectives of the organization.

7 CUSTOMER NEEDS IDENTIFICATION

7.1 Market Planning an Implementation Requirements

Service, process, and product definition should start with determining customer requirements. It should:

a. Determine the need for a product or service

b. Define the market demand and sector, because doing so is important in determining the grade, quality, price, and timing estimates for the product or service

c. Determine customer requirements by a review of contract or market needs, including an assessment of any unstated expectations or perceptions held by customers

d. Obtain concurrence on pertinent customer requirements clearly and accurately within the organization

e. Define competitive positioning

f. Develop a strategy for producing the product or delivering the service

g. Develop a sales strategy utilizing the appropriate mix of product, price, promotion and distribution for a specific market segment

7.2 Service, Process, and Product Brief

A formal statement or outline of product requirements (e.g., a product brief) should be prepared. The product brief translates customer requirements and expectations into a preliminary set of specifications as the basis for concurrent and subsequent design work. The product brief may include:

a. Performance characteristics (e.g., environmental and usage conditions and reliability)

b. Sensory characteristics (e.g., style, color, taste, smell)

c. Installation, serviceability, maintainability, configuration, or fit

d. Product and process regulatory requirements covering the impact on the environment

e. Applicable standards and statutory regulations

f. Packaging

g. Process definitions for production of the product or providing the service

h. Quality assurance/verification

The brief should be a dynamic document and be updated and constantly reverified during the product, service, or process life cycle.

7.2.1 Market Baseline and Development Release

The results of the brief should be reviewed prior to release by competent individuals to ensure its validity.

7.2.2 Configuration Control

A procedure for controlling the release, change, and use of documents that define the market baseline should be established and maintained.

7.3 Customer Feedback Information

An information monitoring and feedback system should be maintained on a continuous basis. Information pertinent to the quality of a product or service should be analyzed, collated, interpreted, and communicated in accordance with defined procedures.

Such information should be utilized to determine the nature and extent of product or service problems in relation to customer experience and expectations.

The feedback should be utilized to provide clues to possible design changes, new service or product offerings, and appropriate management action.

8 SERVICE, PROCESS, AND PRODUCT DEVELOPMENT

8.1 General

The product/service and process development cycle is initiated concurrent with or after development of the product brief.

The development cycle includes engineering development, supplier selection, and operations (including sales and service) planning.

Because these activities continue after the initial design enters the marketplace, product, operation, supplier improvements, and changes require continuous quality management.

8.2 Design Planning

The quality system should ensure that the development process provides clear and definitive technical data for procurement, the execution of work, the operating conditions of processes, and verification of conformance of products and processes to specification requirements.

Management should establish the design process, specifically assign responsibilities for various design duties to activities inside and/or outside the organization, and ensure that all those who contribute to design are aware of their responsibilities for achieving the goals established in the product brief. For a complex system, this should include:

a. Detailed requirement analysis

b. System engineering

c. Decomposition of the product/service/process into its major parts, which may be developed by separate groups

d. Definition of interfaces among the different groups and product/service/process integration

e. Development of detailed specifications

f. Development of appropriate processes and process controls for key service, process, and product features

g. Development of appropriate service, worker, user, and installation instructions

h. Reliability management (see IEC 300 standard)

i. Software (see ANSI/ASQC 730 or ISO 9000-3)

j. Other pertinent activities

The development team should also give consideration to the requirements relating to:

a. Safety

b. Safeguards against misuse

c. Impact on the environment

d. Energy conservation

e. Disposal

f. Legal and other regulatory requirements

Management should establish time-phased design programs with checkpoints appropriate to the nature of the product. The extent of each phase and the stages at which design reviews or evaluations will take place may depend upon the product's application, its design complexity, the extent of innovation and technology being introduced, the degree of standardization, and similarity with past proven designs.

8.3 Product Testing and Measurement

The methods of measurement and test and the acceptance criteria applied to evaluate the service, process, and product during both the design and production phases should be specified for performance and reliability. Parameters should include the following:

a. Performance target values, tolerances, process capability, and attribute features

b. Acceptance and rejection criteria

c. Test and measurement methods, equipment, bias and precision requirements, and computer software considerations

d. Compatibility of the measuring system (repeatability and reproducibility) as a maximum error ratio to tolerance width

e. Ease of use

8.4 Design Qualification and Validation

8.4.1 Design Risk Assessment

The design process should provide periodic evaluation of the design at significant stages. Such evaluation can take the form of analytical methods, such as FMEA (Failure Mode and Effects Analysis), fault tree analysis, or risk assessment, as well as inspection or test of prototype models and/or actual production samples. The amount and degree of testing should be related to the risks identified in the design plan. Independent evaluation may be employed, as appropriate, to verify original calculations, provide alternative calculations, or perform tests. Adequate numbers of samples should be examined by tests and/or inspection to provide appropriate statistical confidence in the results. The tests should include the following activities:

a. Evaluation of performance, durability, safety, reliability, and maintainability under expected storage and operational conditions

b. Inspections to verify that all design features are as intended and that all authorized design changes have been accomplished and recorded

c. Validation of computer systems and software

The results of all tests and evaluations should be documented regularly throughout the qualification test cycle. Review of test results should include analysis of causes of defects and failures.

8.4.2 Design Verification

Design verification may be undertaken independently or in support of design reviews by applying the following methods:

a. Alternative calculations, made to verify the correctness of the original calculations and analyses

b. Testing (e.g., by model or prototype test—if this method is adopted, the test programs should be clearly defined and the results documented)

c. Independent verification, to verify the correctness of the original calculations and/or other design activities

8.5 Design Review

At the conclusion of each phase of design development, a formal, documented, systematic, and critical review of the design results should be conducted. This should be integrated with a review of overall project progress, including progress in developing production, delivery, and sales processes and progress in meeting budget and schedule.

Participants at each design review should include representatives of organization functions affecting quality. The design review should identify and anticipate problem areas and inadequacies and initiate corrective actions to ensure that the final design and supporting processes meet customer requirements. Customer involvement should be utilized when appropriate.

Details of the design review process are described in the ANSI/ASQC standard (TBD): Guide for Formal Design Review.

8.6 Design Baseline and Operations Release

The results of the final design review should be documented in specifications and drawings that define the design baseline. Where appropriate, this should include description of qualification test units "as built" and modified to correct deficiencies during the qualification test programs for configuration control throughout the production cycle.

The total document package that defines the design baseline should require approval by the groups within the organization and suppliers affected by or contributing to the product. This "approval" constitutes the production release and signifies concurrence that the design can be realized.

8.7 Business and Market Readiness Review

The quality system should provide for a review to determine whether production capability and field support are adequate for the new or redesigned product. Depending upon the type of service, process, or product, the review may cover the following points:

a. Availability and adequacy of installation, operation, maintenance, and repair manuals

b. Existence of an adequate distribution and customer service organization

c. Training of field personnel

d. Availability of spare parts

e. Field trials

f. Certification of the satisfactory completion of qualification tests

g. Physical inspection of early production units and their packaging and labeling

h. Evidence of process capability to meet specification on production equipment

i. Production readiness

8.8 Design Change Control (Configuration Management)

The quality system should provide a procedure for controlling the release, change, and use of documents that define the design baseline (resultant product configuration) and for authorizing the necessary work to be performed to implement changes that may affect the service, product, or process during its life cycle. The procedures should provide for the various necessary approvals, specified points and times for implementing changes, removing obsolete drawings and specifications from work areas, and verification that changes are made at the appointed times and places. These procedures should handle emergency changes necessary to prevent production of nonconforming product. Consideration should be given to instituting formal design reviews and validation testing when the magnitude, complexity, or risk associated with the change warrants such actions.

Where suppliers perform design activities, adherence to design change control should be required.

8.9 Design Requalification

Periodic re-evaluation of product should be performed in order to ensure that the design is still valid with respect to all specified requirements. This should include a review of customer needs and technical specifications in light of field experiences, field performance surveys, or new technology and techniques. The review should also consider process modifications. The quality system should ensure that any production and field experience indicating the need for design change is fed back for analysis. Care should be taken that design changes do not cause product quality degradation and that proposed changes are evaluated for their impact on all product characteristics in the design baseline definition.

8.10 Procurement Planning

8.10.1 General

Purchased services, materials, components, and assemblies directly affect the quality. Quality of services such as calibration and special processes should also be considered. The procurement of purchased supplies should be planned and controlled.

The purchaser should establish a close working relationship and feedback system with each supplier to promote continual quality improvements and to avoid or quickly resolve quality.

This close working relationship and feedback system should provide benefits to the purchaser and the supplier.

The procurement quality process should consider the following elements as a minimum:

a. Requirements for specifications, drawings, and purchase orders

b. Selection of qualified suppliers

c. Agreement on quality assurance requirements

d. Agreement on verification methods

e. Provisions for settlement of quality disputes

f. Receiving inspection plans and controls

g. Qualification requirements

h. Receiving quality records

i. Process capability requirements

j. Agreement on delivery schedule

k. Requirements for data files and data formats

8.10.2 Requirements for Procurement Specifications/Drawings and Purchase Orders

The procuring activity should develop appropriate methods to ensure that the requirements for the supplies are clearly defined, communicated, and, most importantly, are completely understood by the supplier. These methods may include written procedures for the preparation of specifications, drawings, and purchase orders; supplier/purchaser conferences prior to purchase order release; and other methods appropriate for the supplies being procured.

Purchasing documents should contain data clearly describing the product or service ordered. Elements that may be included are as follows:

a. Precise identification of style and grade

b. Inspection instructions and applicable specifications

c. Quality system standard to be applied

Purchasing documents should be reviewed for accuracy and completeness before release.

8.10.3 Selection of Qualified Suppliers

Each supplier should have a demonstrated capability to furnish supplies that can meet all the requirements of the specifications, drawings, and purchase order.

The methods of establishing this capability may include any combination of the following:

a. On-site assessment and evaluation of supplier's capability and/or quality system

b. Evaluation of product samples

c. Evaluation of supplier's process capability

d. Past history with similar services, processes, or products

e. Test results of similar services, processes, or products

f. Published experience of other users

g. Previous history of supplier's performance

8.10.4 Agreement on Quality Assurance

A clear understanding should be developed with the supplier on quality assurance for which the supplier is responsible. The assurance to be provided by the supplier may vary as follows:

a. The purchaser relies on supplier's quality assurance system

b. Submission of specified inspection/test data or process control records with shipments

c. 100% inspection/testing by the supplier

d. Lot acceptance inspection/testing by sampling by the supplier

e. Implementation of a formal quality assurance system as specified by the purchaser

f. Other special techniques appropriate to the circumstances

8.10.5 Agreement on Verification Methods

A clear agreement should be developed with the supplier on the methods by which conformance to purchaser's requirements will be verified. Such agreements may also include the exchange of samples of product, inspection, and test data with the aim of assuring measurement agreement and product quality interpretation. Reaching agreement minimizes difficulties in the interpretation of requirements as well as inspection, test, or sampling methods.

8.10.6 Provisions for Settlement of Disputes

Procedures should be established by which settlement of disputes regarding quality can be reached with suppliers. Provisions should exist for dealing with routine and nonroutine matters to improve communication channels between the purchaser and the supplier on matters affecting quality.

8.11 Operations Planning

Operations should be planned so they proceed under controlled conditions in the specified manner and sequence. Controlled conditions include appropriate controls for materials, production equipment, processes and procedures, computer software, personnel, and associated supplies, utilities, and environments.

Process instructions should be documented to the extent necessary to provide proper evidence of control.

Provisions for common practice that apply throughout the operation facility should be similarly documented and referenced in individual work instructions. These instructions should describe the criteria for determining satisfactory work completion and conformity to specification and standards of good work and/or workmanship. Workmanship standards should be defined to the necessary extent, for example, by written standards, photographs, and/or physical samples.

Verification of the quality status of a product, process, software, material, or environment should be considered as important points in the production sequence to minimize effects of errors and to maximize yields. The use of control charts and statistical sampling procedures and plans are examples of techniques employed to facilitate production/process control.

Verification at each stage should relate directly to finished product specifications or to internal requirements, as appropriate, that reflect customer needs and desires. If verification of characteristics of the process itself is not physically or economically practical or feasible, then verification of the product should be utilized. In all cases, relationships between customer needs and in-process controls, in-process specifications, and final product specifications should be developed, communicated to production and inspection personnel, and documented.

All in-process and final inspections should be planned and specified. Documented test and inspection procedures should be maintained, including the specific equipment to perform such checks and tests, as well as the specified requirement(s) and/or workmanship standard(s) for each quality characteristic to be checked.

Efforts to develop new methods for improving production quality and process capability should be encouraged.

8.12 Process Capability

Production processes should be verified as effective and stable in meeting product or service specifications.

When variables data are available, process capability should be evaluated. Capability indexes can be used to evaluate the capacity of a process to meet specified requirements for identified characteristics.

Operations associated with product or process characteristics that can have a significant effect on product or service quality should be identified.

Verification of production processes should include material, equipment, computer system and software, procedures, and personnel.

8.13 Supplies, Utilities, and Environments

Where important to quality characteristics, auxiliary materials and utilities (such as water, compressed air, electric power, and chemicals used for processing) should be controlled and verified periodically to ensure uniformity of effect on the process.

Where an environment (such as temperature, humidity, and cleanliness) is important to quality, appropriate limits should be specified, controlled, and verified.

9 OPERATIONS

Controls should be in accordance with the quality plan established during the service, process, and product development process. These controls should cover:

a. Production

b. Servicing

c. Installation

d. Timely delivery of the product

e. Other post-development activities

9.1 Material Control

9.1.1 Verification of Quality

All materials and parts should conform to appropriate specifications and quality standards before being introduced into production. However, in determining the amount of test and/or inspection necessary, consideration should be given to cost impact and the effect that substandard material quality will have on production flow.

9.1.2 Traceability

Where in-plant or post-factory traceability of material is important to quality, appropriate identification should be maintained throughout the operation processes to ensure traceability to original material identification and quality status.

9.1.3 Handling, Transportation, and Storage

The handling, transportation, and storage methods of materials and product require proper planning, control, and a documented system for incoming materials, materials in process, and finished goods; this applies not only during delivery but up to initial customer use.

The method of handling and storage of materials should provide for the correct pallets, containers, conveyors, and vehicles to prevent damage due to vibration, shock, abrasion, corrosion, temperature, or any other conditions occurring during handling, transportation, and storage. Items in storage should be checked periodically to detect possible deterioration.

9.1.4 Special Protection

Items requiring special protection during any phase of operation (factory use, transportation, servicing, etc.) should be identified, and procedures should be maintained and provided to all personnel that may be impacted. This would include designers, factory personnel, distribution personnel, and customers.

9.1.5 Identification

The marking and labeling of materials should be legible, durable, and in accordance with specifications. Identifications should be provided to prevent mix-ups.

Final product marking should be durable to identify a particular product by all customers and distributors in the product chain if a recall or special inspection becomes necessary.

9.1.6 Packaging

The methods of cleaning, preserving, and protecting the product throughout its intended handling, storage, and transportation cycle to the ultimate customer receipt should be detailed in written instructions, as appropriate.

9.1.7 Spare Parts

Assurance should be provided for an adequate logistic backup, to include technical advice, spares or parts supply, and competent servicing. Responsibility should be clearly assigned and agreed upon among suppliers, distributors, and users.

9.2 Equipment Control and Maintenance

All production equipment, including fixed machinery, jigs, fixtures, tooling, templates, patterns, and gauges, should be proved for precision and absence of bias prior to use. Special attention should be paid to computers used in controlling processes, and especially the maintenance of the related software.

Equipment should be appropriately stored and adequately protected between use and verified or calibrated at appropriate intervals to ensure control of bias and precision.

A program of preventive maintenance should be established to ensure continuing process capability. Special attention should be given to equipment characteristics that contribute to key product quality characteristics.

Tools or equipment that are introduced for servicing products during or after installation should have their design and function validated, as for any new product.

9.3 Measurement Process Control

9.3.1 Measurement Control

Sufficient control should be maintained over all measurement systems used in the development, manufacture, installation, and servicing of product to provide confidence in decisions or actions based on measurement data.

Control should be exercised over all elements of the measurement process, including instruments, operating personnel, measurement methods, environmental conditions, and data collection methods.

Operations fixtures and process instrumentation that can affect the specified characteristics of a product, process, or service should be suitably controlled.

Procedures should be established to monitor and maintain the measurement process itself under statistical control, including equipment, procedures, and operator skills.

Measurement process uncertainties should be compared with the requirements and appropriate action taken when precision and/or bias requirements are not achieved.

9.3.2 Factors of Control

The control of the measurement process should include the following factors, as appropriate:

a. Correct specification and acquisition, including range, bias, precision, robustness, and durability under specified environmental conditions for the intended service

b. Initial calibration validation of the required bias limits and precision prior to first use

c. The software, and procedures controlling automatic processes equipment, should be tested

d. Periodic adjustment, repair, and calibration, considering manufacturer's specification, the results of prior calibration, and the method and extent of use, to maintain the required accuracy in use

e. Documentary evidence covering identification of instruments, frequency of recalibration, calibration status, and procedures for recall, handling, storage, adjustment, repair, calibration, installation, and use

f. Traceability to reference standards of known accuracy and stability, preferably to national or international standards or, in industries or products where such do not exist, to specially developed criteria

9.3.3 Measurement Controls for Outside Organizations

The control of measuring and test equipment and procedures extends throughout the product life cycle chain (suppliers, installers, or any others performing measurement).

The facilities of outside organizations that are used for measurement, testing, or calibration services should comply with these requirements.

9.3.4 Corrective Action

Where measuring processes are found to be out of control or where measuring and test equipment is found to be outside the required calibration limits, corrective action is necessary.

An evaluation should be made to determine the effects on completed work and to what extent reprocessing, retesting, calibration, or complete rejection may be necessary. In addition, investigation of cause is important in order to avoid recurrence. This may include review of calibration methods and frequency, training, and adequacy of test equipment.

9.4 Process Control

Consideration should be given to production processes in which process control is particularly important to service, process, and product quality. Consideration should be required for service, process, or product characteristics that are not easily or economically measured; for special skills required in their operation or maintenance; or for a service, process, or product whose results cannot be fully verified by subsequent inspection and test. More frequent verification should be made of:

a. The accuracy and variability of equipment used to make or measure product, including settings and adjustments

b. The skill, capability, and knowledge of operators to meet quality requirements

c. Special environments, time, temperature, or other factors affecting quality

d. Certification records maintained for personnel, processes, and equipment, as appropriate

e. That any changes to service, product, or process parameters or their interactions are quickly detected and corrections implemented

f. Ongoing supplier monitoring

9.5 Documentation

Requirements, process instructions, specifications, and drawings should be controlled as specified by the quality system.

This information should be readily available to the appropriate personnel.

9.5.1 Installation

Instructional documents should clearly delineate proper installations and should include provisions which preclude improper installation or factors degrading the quality, reliability, safety, and performance of any product or material.

9.5.2 After-Sales Servicing

Instructions for the assembly and installation, commissioning, operation, care and maintenance, safety, warranty, spares or parts lists, and servicing of any product should be comprehensive and supplied in a timely manner. The suitability of instructions for the intended reader should be verified.

9.6 Process Change Control

Process change authority should be clearly designated and, where necessary, customer approval should be sought. All changes to processes should be documented. The process change implementation should be covered by defined procedures.

A service, process, or product should be evaluated after any change to verify that the change instituted had the desired effect upon quality. Any changes in the relationships between process and product characteristics resulting from the change should be documented, communicated, and appropriately controlled.

9.7 Verification

9.7.1 Incoming Materials and Parts

The method used to ensure quality of purchased materials, component parts, and assemblies that are received into the production facility will depend on the importance of the item to quality, the state of control and information available from the supplier, and impact on costs.

9.7.2 Completed Verification

To augment inspections and tests made during production, two forms of final verification are available. Either or both of the following may be used, as appropriate:

a. Acceptance inspections or tests may be used to ensure that items or lots produced have met performance and other quality requirements. Examples include screening (100% of items), lot sampling, customer acceptance of service performed, and continuous sampling.

b. Periodic or continuous quality auditing of representative samples.

9.7.3 Control of Verification Status

Verification status of material and assemblies should be identified throughout production. Such identification may take the form of stamps, tags, work orders, or inspection records.

The identification should include the ability to distinguish between verified and unverified material and indication of acceptance at the point of verification. It should also provide traceability to the unit responsible for the operation.

9.8 Nonconformity

Provision should be made for the managing and control of nonconforming services, processes, or products.

9.8.1 Analysis

There should be an immediate assessment of the risk associated with the nonconformity. If a product is involved, the material should be segregated.

9.8.2 Reporting

The nonconformity should be reported to the appropriate people that may be impacted by the nonconformity.

9.8.3 Initial Response

A plan of action should be established and implemented to remedy the nonconformity impact on the customer and operations.

9.8.4 Review

A review should be performed to determine if corrective action is required. The results of the review should be documented.

The significance of a problem's effect on quality should be evaluated in terms of its potential impact on such aspects as customer satisfaction, performance, reliability, and safety, as well as operation and quality costs.

9.9 Corrective Action

Corrective action is initiated in parallel with nonconformity review.

9.9.1 Assignment of Responsibility

The responsibility and authority for instituting corrective action should be defined. The coordination, recording, and monitoring of corrective action related to all aspects of the organization or a particular product or service should be assigned to a particular function within the organization. However, the analysis and execution may involve a variety of functions, such as sales, design, production engineering, production, and quality.

9.9.2 Investigation of Possible Causes

The relationship of cause and effect should be determined, with all potential causes considered. Important variables affecting the capability of the process to meet required standards should be identified.

9.9.3 Analysis of Problem

In the analysis of a quality-related problem, the root cause should be determined before the preventive measures are planned. Careful analysis of the product or service specifications and of all related processes, operations, quality records, service reports, and customer complaints should be conducted. Statistical methods can be useful in problem analysis.

9.9.4 Preventive Action

Actions should be taken to prevent a future recurrence of a nonconformity. These actions may change a process, packaging, transit, or storage procedure; revise a product specification; and/or revise the quality system.

Preventive action should be initiated to a degree appropriate to the magnitude of potential problems.

9.9.5 Process Controls

Sufficient control of processes and procedures should be implemented to prevent recurrence of the problem. When the preventive measures are implemented, their effect should be monitored in order to ensure that desired goals are met.

9.9.6 Disposition of Nonconforming Items

If the initial response has not resolved work in progress, remedial action should be instituted as soon as practical.

9.9.7 Recovery Plan for Product or Services Sold

The recovery plan should address:

1. The recall of services sold or product sold (including products in finished goods warehouse and in transit). Recall decisions are affected by considerations of safety, liability, and customer satisfaction.
2. Other actions taken on the total customer base to maintain customer satisfaction.

9.9.8 Permanent Changes

Permanent changes resulting from corrective action should be recorded in work instructions, operations processes, product specifications, and/or the quality system and should be communicated to all affected personnel. It may also be necessary to revise the procedures used to detect and eliminate potential problems.

9.10 Marketing Reporting and Product Supervision

An early warning system may be established for reporting instances of product failure or shortcomings, as appropriate, particularly for newly introduced products, to ensure rapid corrective action.

A feedback system regarding performance in use should exist to monitor the quality characteristics of the product throughout its life cycle. This system should be designed to analyze, as a continuing operation, the degree to which the product or service satisfies customer expectations on quality, including safety and reliability.

Information on complaints, the occurrence and modes of failure, customer needs and expectations, or any problem encountered in use should be made available for design review and corrective action.

10 QUALITY IMPROVEMENT

10.1 Purpose

A process should be developed, implemented, and documented to involve everyone in the continuous improvement of the quality of the business processes, products, and services.

The quality improvement process should focus on improving customer satisfaction and the effectiveness and efficiency of business processes.

This process should be proactive to achieve improvement, rather than a reaction to eliminate problems.

10.2 Plans and Goals

Quality improvement goals should be established. They should be closely integrated with the overall business goals and provide focus for increasing customer satisfaction and process efficiency.

Quality improvement plans and activities should be based upon the result obtained from the quality information system (quality audits, quality data evaluation, purchaser feedback, etc.).

Goals should be clearly understandable, challenging, pertinent, and agreed to by all who work together to achieve them.

10.3 Communication and Teamwork

Open communication and teamwork should be extended throughout the whole supply chain, including suppliers and customers. Organizational and personal barriers should be removed and eliminated.

10.4 Organizing for Quality Improvement

The organization for quality improvement should identify opportunities both vertically within the organizational hierarchy and horizontally in the processes that flow across organizational boundaries.

Improvements should come from people working within the process and from people external to the process.

10.4.1 Within Organization Function

Within an organizational function, responsibilities include:

- The management processes of the organization, such as defining the mission of the organization, clarifying roles and responsibilities, acquiring and assigning resources, providing education, strategic planning, and recognition

- The work processes of the organization

- The measurement and tracking of the performance of the organization's processes

- The administrative support processes, such as secretarial, budgeting, and purchasing

- Building and maintaining an environment that empowers, enables, and charges members of the organization with continuous improvement of quality

10.4.2 Cross Functional

Within processes that flow across organizational boundaries, responsibilities include:

- Defining and agreeing on the purpose of each process and its relationship with the objectives of the organization

- Establishing and maintaining communication among departments

- Identifying both internal and external customers of the process and determining their needs and expectations

- Identifying suppliers to the process and communicating to them their customer needs and expectations

- Searching for process improvement opportunities, allocating resources for improvement, and overseeing implementation of these improvements

10.4.3 Criteria

An organization should address the following:

- A means for providing policy, strategy, goals, guidance, support, and broad coordination of the organization's quality improvement activities

- A means of identifying cross-functional quality improvement needs and goals and assigning resources to pursue them

- A means to pursue quality improvement goals by team projects (within direct organizational responsibilities and cross functional) and individual efforts

- A means for encouraging every member of the organization to pursue quality improvement activities related to their work and for coordinating these activities

- A means to document the results of quality improvement projects should be established

- A means to evaluate the system's effectiveness

10.5 Methodology

10.5.1 General

An organization should be motivated and managed for quality improvement.

An organization should continuously undertake quality improvement projects of varied complexity.

Quality improvement project results should show an accumulation of results over time.

Quality improvement projects should consist of a consistent, disciplined series of steps based on data collection and analysis.

10.5.2 Initiating a Project

The need, scope, and importance of a project should be clearly defined and demonstrated by providing the relevant background in quantitative terms. A person or a team should be assigned to work on the project. A schedule should be established and provisions for allocation of resources be made. Provision for periodic management reviews should be made.

10.5.3 Investigating Possible Causes

The project should attempt to objectively investigate possible causes of the problem.

10.5.4 Establishing Cause-and-Effect Relationships

Data should be analyzed to formulate possible cause-and-effect relationships. The relationships that have a high degree of consistency with the data need to be tested.

10.5.5 Taking Action

After cause-and-effect relationships have been established, actions to prevent recurrence of the problem should be developed and evaluated. The appropriate plan of action should be developed and implemented.

10.5.6 Follow-Up

After implementing the solution, appropriate data must be collected and analyzed to confirm that an improvement has been made. The data should be collected on the same basis as prior to the implementation. Investigations for other side effects, either desirable or undesirable, should be made.

10.5.7 Sustaining the Gains

The improved process needs to be controlled at the new performance level. This usually involves a change of specifications and/or operating or administrative procedures and practices, necessary education and training, and making sure that these changes become an integral part of the job content of everyone concerned.

10.6 Supporting Tools and Techniques

Use of various problem-solving tools should support the quality improvement process as specified in paragraph 5.4.2. Other tools may be used as appropriate.

INDEX

C

Calibration, 97, 150
Canada, 15
Career planning, 133
Cause-and-effect diagram, 56, 57, 128
CEN, see European Committee for Standardization
CENLEC, see European Committee for Electrochemical Standardization
Certainty, 32
Certification
  in EC, 8
  of registrars, 15–16
Characteristic parameters, for service, 77–80
Communication, 129, 158
Configuration management, 141
Conformance, 50
Constancy of purpose, 37, 39
Consultants, 21–22
Continuous improvement, 65, 123
Contract review, 7, 74, 75, 97
Control chart, 128
Control mechanism, 123
Control of nonconforming product, 7, 75, 83, 89, 97, 98
Corrective action, 7, 33, 75, 83, 89, 97, 98, 151, 154–156
Cost of quality, 27–36, 52, 114, 125–127
  types of, 126–127
Craft society, quality in, 23–24
Critical risk, defined, 76
Crosby, Philip, 27, 28, 29
  make-certain program, 35
  management style evaluation, 36
  quality improvement program, 32–34
Culture, management leadership and, 116
Customer expectations, 110

Customer feedback, 116, 135–136
Customer requirements, 45–46, 48, 49, 50, 56, 65, 101
  defining for service, 74
  identification, 134–136
Customer satisfaction, 65, 101, 124–125
Customer service, defined, 73

D

Decision-makers, 68, 69, 70
Delivery, see Handling, storage, packaging, and delivery
Deming, W. Edwards, 26, 37–41
Deming Prize, 27
Denmark, 10, 12
Deployment, 51
Design, 43–44, 50, see also Product design; Quality design
Design change, 141
Design control, 7, 75, 89, 97
Design of experiments, 44, 48–49, 80, 128
Design planning, 136–137
Design process, 136–141
Design requalification, 141
Design review, 139–140
Direct labor costs, 95, 96
Direct material costs, 95, 96
Disputes, 144
Documentation, 118–119, 152
Document control, 7, 75, 97, 119
Document retention, 120–121
Dodge, Harold, 25
Dutch Council for Accreditation, 16

E

EAC, see European Accreditation of Certification
EC, see European Community

DATE DUE

JUL 1 1 1997

MAR 1 5 1998

JAN 0 8 2000

MAY 3 0 2014

DEMCO, INC. 38-2971